Springer Theses

Recognizing Outstanding Ph.D. Research

Aims and Scope

The series "Springer Theses" brings together a selection of the very best Ph.D. theses from around the world and across the physical sciences. Each volume has been selected for its scientific excellence by an internationally renowned institute or department. For greater accessibility to non-specialists, the published versions include an extended introduction, as well as a foreword by the student's supervisor explaining the special relevance of the work for the field. The series as a whole provides a valuable resource both for newcomers to the research fields, and for other scientists seeking detailed background information on special questions. Finally, it serves as an accredited documentation of the valuable contributions made by today's younger generation of scientists.

How to Apply

Theses prepared at internationally leading universities and institutes may be nominated for publication by supervisors with the endorsement of the head of department. Each such department is entitled to nominate one thesis per year. Enquiries and requests for a nomination form should be sent, together with the thesis, to Ms. Nathalie Berg (see link on right).

Eligibility Criteria

To qualify for consideration, theses should fulfil the following criteria:

- They must be written in good English.
- The topic should fall within the confines of Chemistry, Physics, Earth Sciences, Engineering and related interdisciplinary fields such as Materials, Nanoscience, Chemical Engineering, Complex Systems and Biophysics.
- The work reported in the thesis must represent a significant scientific advance.
- A maximum of 30% of the text may be a verbatim reproduction from the author's previous publications.
- They must have been examined and passed during the 12 months prior to nomination.
- The theses should have a clearly defined structure including an introduction accessible to new PhD students and scientists not expert in the relevant field.
- After acceptance, theses should be supplemented by a "Supervisor's Foreword" outlining the wider scientific relevance of the work.

Indexed by zbMATH.

Jeonghwan Park

5 T High Temperature Superconductor 3 Pole Wavelength Shifter Magnet for Accelerator-based Light Sources

Doctoral Thesis accepted by
the Seoul National University, Seoul, Korea (Republic of)

Springer

Author
Dr. Jeonghwan Park
Department of Electrical and Computer Engineering, College of Engineering
Seoul National University
Seoul, Korea (Republic of)

Supervisor
Dr. Seungyong Hahn
Department of Electrical and Computer Engineering
Seoul National University
Seoul, Korea (Republic of)

ISSN 2190-5053 ISSN 2190-5061 (electronic)
Springer Theses
ISBN 978-981-96-8678-0 ISBN 978-981-96-8679-7 (eBook)
https://doi.org/10.1007/978-981-96-8679-7

This Springer imprint is published by the registered company Springer Nature Singapore Pte Ltd.
The registered company address is: 152 Beach Road, #21-01/04 Gateway East, Singapore 189721, Singapore

Supervisor's Foreword

This dissertation presents design, fabrication, and operation of the first high-temperature superconducting (HTS) magnet for a wavelength shifter, achieving a record high magnetic field of 5 tesla. The work was carried out by a team under my supervision at the Applied Superconductivity Laboratory, Seoul National University during 2022–2024 in close collaboration with Pohang Accelerator Laboratory (led by Dr. Garam Hahn and Dr. Seunghwan Shin) and Kangwon National University (led by Dr. Seyong Choi). Dr. Jeonghwan Park played the leading role in the team over the entire R&D cycle that includes: (1) electromagnetic, mechanical and thermal designs of the magnet as well as auxiliary parts including mechanical support and cryogenic system; (2) hands-on manufacturing of the magnet and assembly using in-house machines; (3) operation under a conduction-cooled cryogenic condition followed by in-depth analyses with advanced simulation tools; (4) multi-institute-and-big-group project management with outstanding leadership and documentation including data organization and paper writing. All of these achievements undoubtedly advance the high temperature superconductor magnet technology in general as well as the accelerator application in particular.

The 5 T wavelength shifter HTS magnet developed in this study is expected to drastically enhance the performance of synchrotron light sources. When applied to the Pohang Light Source II (PLS II), for example, the new technology may generate over 1000 times brighter light in the photon energy range of 100 keV, significantly enhancing its capabilities for high-resolution imaging and advanced material analysis. Furthermore, this research was conducted as a part of the National Synchrotron Light Source Localization Project, which successfully localized key equipment for synchrotron accelerators in the domain of Korean community. And, the results align with and will contribute to the ongoing construction of the next-generation synchrotron accelerator in Korea (2021–2029, a total budget of 1.2 trillion KRW) as a powerful new insertion device that is currently unavailable even with the state-of-the-art magnet technology. Currently, the Pohang Accelerator Laboratory has established a new cryogenic superconducting research facility dedicated to the HTS magnet R&D for particle accelerator, demonstrating its impact on both domestic and global research landscapes.

Beyond its contributions to HTS magnet technology, this study also aims for another important milestone: the first operation of an HTS magnet in a real-world light source. Given these pivotal achievements, efforts are currently underway to install this 5 T wavelength shifter magnet in the PLS-II accelerator at the Pohang Accelerator Laboratory. Obviously, this will face a whole new level of technical challenges but, once completed, the outcomes would expand the horizon of accelerator technology, not limited to magnet, and enhance capabilities in diverse fields such as particle physics, cancer therapy, discovery of advanced materials, new drug development, and more. Given its potential as a valuable reference for high field accelerator magnet research, I highly recommend this dissertation for publication in Springer Nature to maximize its impact on our community.

Last but not least, we would like to thank our international collaborators including CERN (European Organization for Nuclear Research), INFN-LASA (Istituto Nazionale di Fisica Nucleare-Laboratorio di Acceleratori e Superconduttivita Applicata), and Argonne National Laboratory, for their contributions to HTS accelerator magnet R&D. It is our aim that the achievements reported in this dissertation will serve as a valuable reference for ongoing and future collaborative research efforts to build next-generation high field HTS magnets for a variety of accelerator applications.

Seoul, Korea (Republic of)
March 2025

Dr. Seungyong Hahn

Abstract

In this paper, we developed the world's first 5 T conduction-cooled metal-insulated (MI) wavelength shifter (WLS) magnet using high-temperature superconductor (HTS) technology. The main scope of this paper is to cover the overall development process, including design, fabrication, and operation, of conduction-cooled MI HTS WLS magnet. First, a design method for a conduction-cooled MI WLS magnet considering electromagnetic, mechanical, thermal, and operating characteristics was proposed. Next, core manufacturing technology was developed for the development of conduction-cooled MI HTS WLS magnet. In addition, to predict the magnetic field operation characteristics of HTS WLS magnet, a three-dimensional screening current induced magnetic field analysis model considering the nonlinear characteristics of iron was developed.

An accelerator-based light source is a device that produces high-flux photons using a particle accelerator. When using a high-field WLS magnet, high-flux photons in a wider energy spectrum are obtained, which means that the generated photons can be used in a wider variety of experiments. With the potential that a high magnetic field can be generated with high-temperature superconductor (HTS) technology, HTS accelerator magnets are being studied by many researchers. However, HTS accelerator magnets are yet to be applied in the actual accelerator system due to difficulties such as Lorentz force-induced high mechanical stress, absence of design and manufacturing technologies of HTS accelerator magnets, and field distortion due to the screening current-induced field. Difficulties with HTS magnet protection against the quench and the high cost of liquid helium are additional challenges for HTS magnets. Here, we suggest the concept of a liquid helium-free conduction-cooled metal-insulated (MI) HTS magnet, and we discuss the research process to resolve the aforementioned challenges.

This thesis presents the overall development process of the MI HTS magnet including design, fabrication, and operation of the suggested MI HTS WLS magnet. The design methods of the conduction-cooled MI HTS WLS magnet are suggested. First, three dimensional electromagnetic design is performed considering field strength and uniformity using multi-width method. Next, a new structure to reinforce the straight section of the racetrack winding was suggested and the mechanical analysis model considering the suggested structure was developed. In addition, a design was conducted to limit the maximum temperature of the magnet under 20 K considering cryogenic heat load and thermal conduction path. Finally, the charging and discharging delay effect of the MI WLS magnet was analyzed using the lumped circuit equivalent model.

Next, this thesis suggests core manufacturing technologies for the MI HTS WLS magnet development. the overall manufacturing process including winding, joint, lead, structure, cryogenics, and operation/evaluation was defined. In particular, an analysis model to estimate winding dimension change by the winding tension and design methods of the suggested preload structure were developed. With the suggested manufacturing technologies, we demonstrated the conduction-cooled MI HTS magnet.

Finally, we developed the 3D screening current-induced field analysis model considering the nonlinear properties of iron. Screening current distributions are analyzed by the H-formulation method considring the nonlinear BH characteristics of iron core, and screening current induced field is calculated by the segmentation method. The simulation model is verified by comparing the calculation results with the tested magnetic field. The effect of screening current-induced field in beam operation is also discussed, and we predicted that the screening current induced field can be mitigated by the additional sidepole current.

The developed HTS magnet is the first WLS magnet with the HTS technology. One of the significance of this study is the achievement of the highest magnetic field of 5 T for HTS accelerator magnets, where the current record field was 4.52 T by CERN. The developed design and manufacturing technologies may be applied to develop future high-field accelerator magnets. Another impact of this study is that this is the first domestically produced HTS WLS magnet. It is anticipated to increase the performance of the accelerator-based light sources, where the 1000 times brighter photon can be obtained at 100 keV photon energy ranges when the developed magnet is applied to the Pohang Light Source II storage ring.

Competing Interests The author has no competing interests to declare that are relevant to the content of this manuscript.

Contents

Nomenclature

α	Thermal contraction (1/K)
β	Speed of charged particle divided by speed of light
$\Delta_{z_i \in c,m}$	Deformation by thermal contraction and magnetic force (A)
$\dot{N}$	Radiated photon flux (photons/s)
ϵ	Beam energy (eV)
ϵ_0	Electric permeability (8.854×10^{-12} F/m)
ϵ_r	Radiation emissivity
η_g	Constant of proportionality for convection heat transfer (W/m^2·Pa·K)
γ	Beam energy divided by rest mass energy
λ	Penetration depth (m)
λ_u	Magnetic period length of undulator field (m)
μ_n	Friction coefficient bolt head
μ_t	Friction coefficient bolt thread
v	Poisson's ratio
ω_c	Critical photon frequency (Hz)
ϕ	Thread angle of bolt (deg)
$\sigma_{i \in (x,y,x',y')}$	Root mean square of beam size (mm) and beam divergence (rad)
θ	Field angle with respect to c axis of superconductor film (deg)
ξ	Coherence length (m)
A	Magnetic vector potential (J/A·m)
A_g	Gain of the data acquisition system
B	Magnetic field intensity (T)
b	Bit number of the data acquisition system
B_{meas}	Measured magnetic field (T)
B_M	Main pole field (T)
B_p	Brighness of photon beam (ph/s·m^2·rad^2·0.1%BW)
B_{scf}	Magnetic field calculated considering screening current distributions (T)
B_{unf}	Magnetic field calculated considering uniform current distribution (T)
C	Heat capacity (J/K)
c	Speed of light in a vacuum (3×10^8 m/s)

E	Electric field (V/m)
e	Charge of electron
E_c	Critical photon energy
$E_{i \in r,\theta,z}$	Young's modulus in radial, hoop, and axial direction (GPa)
F	Lorentz force (N)
g	Magnetic gap (m)
h	Planck constant (6.626×10^{-34} J·s)
H_c	Critical magnetic field intensity of superconductor (A/m)
I	Current (A)
I_θ	Azimuthal current (A)
I_{beam}	Beam current (A)
I_c	Critical current of superconductor (A)
I_r	Leakage current (A)
J_c	Critical magnetic field intensity of superconductor (A/m^2)
k	Cold end temperature (K)
k	Spring constant (N/m)
k	Thermal conductivity (W/m)
L	Inductance (H)
l_M	Straight section length of main pole racetrack shape (m)
l_{REBCO}	Usage of REBCO conductor (equivalent to 4 mm tape) (m)
l_S	Straight section length of side pole racetrack shape (m)
m_0	Rest mass (kg)
n	Index of power law
N_i	Number of multi layer insulation
N_M	Number of main pole turns
N_{stack}	Number of main pole stacks
N_S	Number of side pole turns
p	Momentum of particle (kg·m/s)
P_g	Pressure (Pa)
P_s	Radiated power of synchrotron radiation (W)
p_w	Twist pitch of bolt (m)
Q	Heat rate (W)
q	Charge of particle
q_c	Conduction heat flux (W/m^2)
q_g	Convective heat flux (W/m^2)
q_r	Radiation heat flux (W/m^2)
R	Bending radius deflected beam (m)
R_{ct}	Contact resistivity (Ω·cm^2)
R_c	Thermal resistance (K/W)
r_{DAQ}	Resolution of the data acquisition system
r_M	Straight section curvature of main pole racetrack shape (m)
r_n	Radius of bolt head (m)
R_{sc}	Superconductor resistivity (Ω)
r_S	Straight section curvature of side pole racetrack shape (m)
r_t	Radius of bolt thread (m)

T	Temperature (K)
T	Winding tension (Pa)
T_c	Critical temperature of superconductor (K)
T_w	Warm end temperature (K)
u	Displacement (m)
V	Voltage (V)
V	Volume (m^3)
v	Velocity of beam (m/s)
V_{lim}	Voltage range of the data acquisition system
w	Width of conductor (m)
$W_{i \in x,y,z}$	Dimension of cover iron (m)
W_m	Magnetic energy (W)
x	Transverse direction coordinate
x'	Deflection of beam (rad)
y	Gravitational direction coordinate
z	Longitudinal direction coordinate
F_p	Bolt preload force (N)
$h_{i \in mag,LN_2}$	Entalphy of magnet or liquid nitrogen (J)
T_p	Bolt fastening torque (N·m)

List of Figures

List of Tables

Chapter 1
Introduction

1.1 Research Motivation

1.1.1 High-Field Electromagnets for Accelerator-Based Light Sources

Particle accelerator uses electromagnetic force to accelerate and manipulate the orbit of charged particles. Especially, the role of the magnetic field is the deflection of high-energy charged particles in Eq. (1.1), and a stronger magnetic field enables a larger deflection of charged particles.

$$\vec{F} = q(\vec{E} + \vec{v} \times \vec{B}) \tag{1.1}$$

$\vec{F}$, $\vec{E}$, q, $\vec{v}$, and $\vec{B}$ are electromagnetic force, electric field, charge of particle, velocity of particle, and magnetic field. Electromagnets are key components of particle accelerators to produce the magnetic field. Particularly, superconducting accelerator magnets are a desirable option for electromagnets in terms of magnetic field, since they can produce a high magnetic field of > 2 T, which is a conventional limit of normal conducting electromagnets.

Requirement of High Field for Accelerator-Based Light Sources

Accelerator-based light sources are particle accelerators that generate a high-flux photon beam. Figure 1.1 shows an example of the accelerator-based light source [1]. Magnet is used to confine the electrons in a circular orbit while generating a high-flux photon beam. Beamline is an experimental site to utilize high-flux photon beams for dedicated experimental purposes such as life science, material science, nanotechnology, chemistry, and many other scientific applications.

J. Park, *5 T High Temperature Superconductor 3 Pole Wavelength Shifter Magnet for Accelerator-based Light Sources*, Springer Theses,
https://doi.org/10.1007/978-981-96-8679-7_1

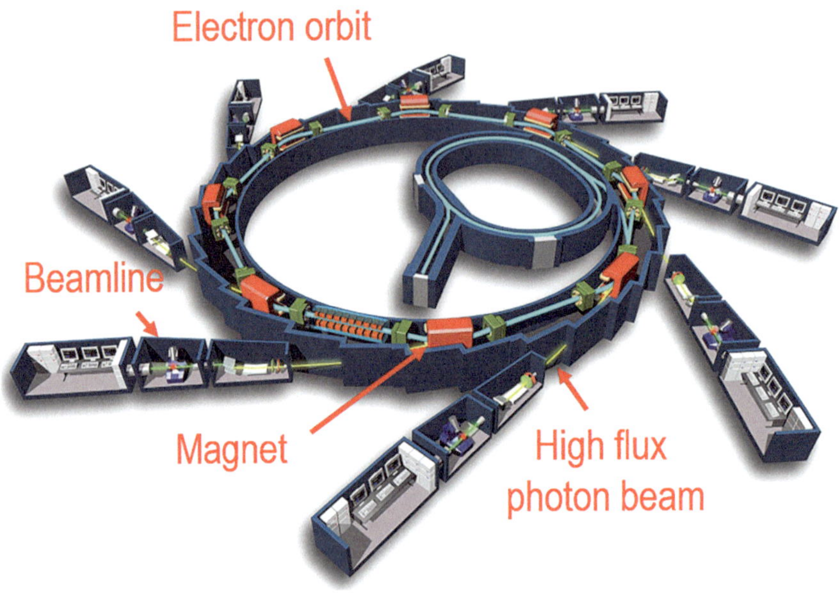

Fig. 1.1 An example of accelerator-based light sources [1]

The basic principle to generate high-flux photon beams for accelerator-based light sources is synchrotron radiation, electromagnetic radiation by deflection of the high-energy electron beam as shown in Fig. 1.2 [2]. The radiated photon density has a wide spectrum in photon energy, and critical energy (E_c) of the radiated photon, which is proportional to magnetic field, is represented by (1.2) [3].

Fig. 1.2 Synchrotron radiation emitted by deflecting electron bunch using bending magnets [2]

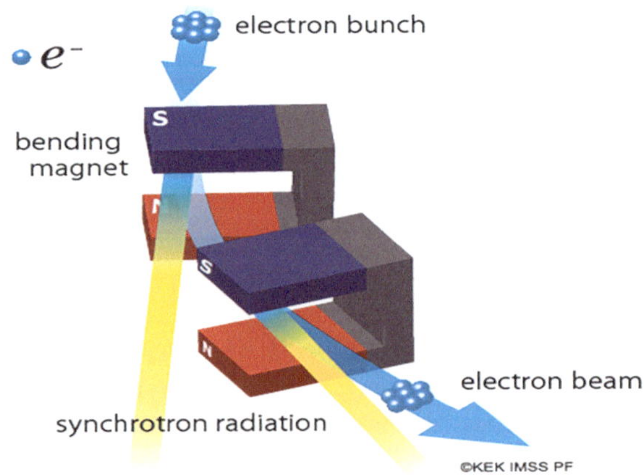

$$E_c = \frac{3hq\gamma^2}{4\pi m\beta} B \tag{1.2}$$

where h, q, γ, B, m, and β are Planck's constant, a charge of the particle, beam energy divided by rest mass energy, magnetic field, the mass of the particle, and speed of beam divided by speed of light. Recent accelerator-based light sources use specially designed magnets to generate high-flux photon beams. These magnets are called insertion devices since they are inserted in a dedicated track of the accelerator. Insertion devices with high magnetic field may result in the generation of high-flux photon beams at higher photon energy ranges.

Representative experiment techniques using high-flux photon beams may be diffraction and scattering, spectroscopy, and imaging. Required photon beam characteristics vary depending on experimental purposes. High-energy (>40 keV) photons have unique advantages over conventional photon sources, such as high penetration depth and small diffraction angles. Figure 1.3 shows the penetration depth depending on the photon beam energy. High-energy photon beam enables imaging of thick materials in a non-destructive way owing to its deep penetration depth. Imaging with low-energy photon beam requires slicing of the sample, therefore the original sample may be damaged. Figure 1.4 shows an example of high-energy photon beam used for paleontology to scan burrow cast with 3D tomographic microscopy [5]. Additionally, high-energy photons can also be used in imaging, paleontology, material studies, and many other scientific experiments [5–10].

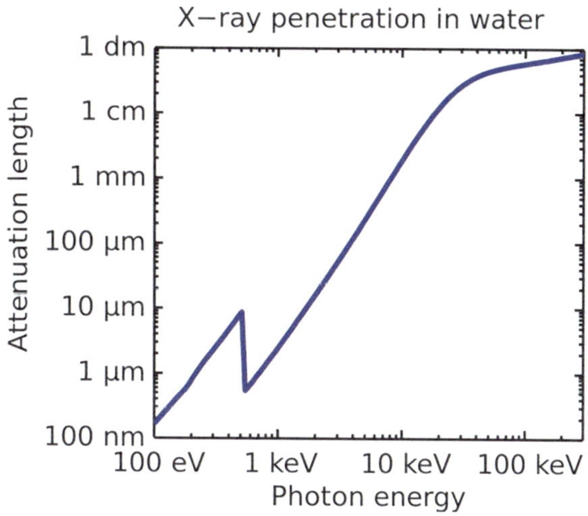

Fig. 1.3 Penetration depth of photon beam depending on its energy [4]

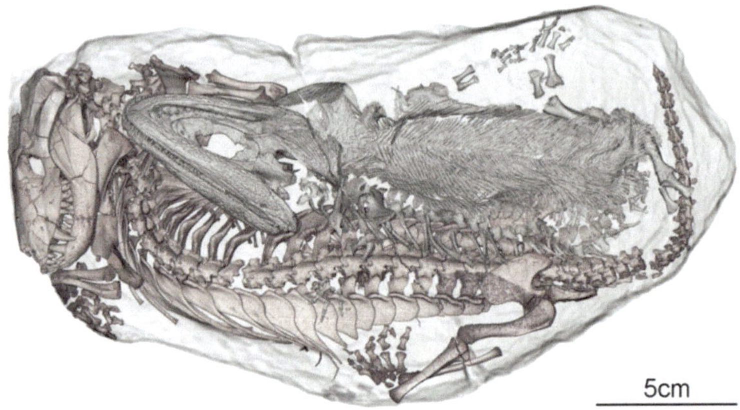

Fig. 1.4 High-energy photon beam used to scan burrow cast with 3D tomographic microscopy [5]

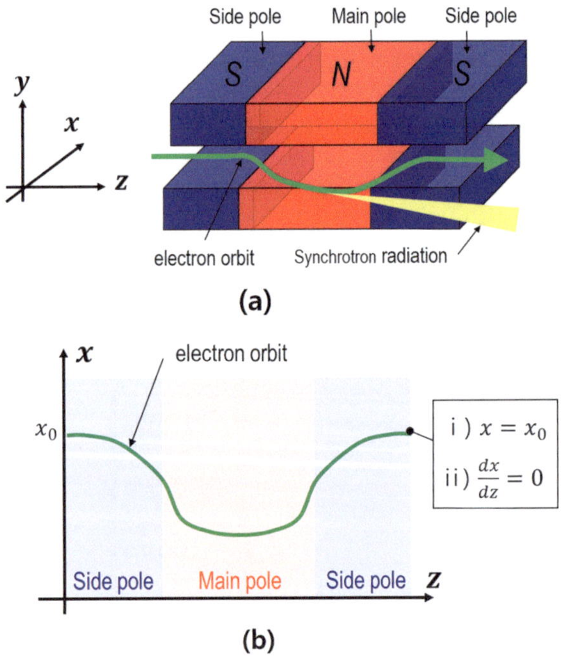

Fig. 1.5 **a** Typical topology of wavelength shifter magnet; **b** electron orbit in the x-z-plane

1.1.2 *Major Research Projects for Accelerator-Based Light Sources*

National Projects to Develop Accelerator-Based Light Sources

Figure 1.6 shows a view map of the Korean 4th-Generation Storage Ring (Korea-4GSR) and Pohang Light Source-II (PLS-II). The PLS-II is a third-generation light source in Korea that started user operation in 2012. Table 1.1 shows the specification of copper or permanent magnet-based electromagnets in PLS-II. Superconducting magnets are not used in PLS-II. In 2021, a new domestic project has also been embarked on building a new light source, called Korea-4GSR [11]. As the domestic large-scale accelerator project to install the next-generation accelerator light source has begun, the demand for high-performance magnets for them is increasing [12–22].

Fig. 1.6 Selected accelerator-based light sources in Korea: PLS-II in operation and 4GSR to be constructed

Table 1.1 Specification of electromagnets used in PLS-II

Magnet type	Field [T]	Period length [mm]	Magnet length
In-vacuum undulator	0.97	20	1.35 or 1.8
Multipole wiggler	1.80	100	2.0
Multipole wiggler	2.02	140	1.12
Out vacuum undulator	0.90	68	3.06
Elliptic undulator	0.684	58	3.20
Elliptic undulator	0.79	72	2.58
Elliptic undulator	0.88	114	3.53
In-vacuum undulator	0.606	10	1.02

International Projects for Developing High-Performance Insertion Devices

- **Permanent magnet-based insertion devices**: Permanent magnets are widely used for undulator-type insertion devices, where relatively low magnetic fields are required [27]. Recently, cryogenic permanent magnets operating at around 77 K have been used to further increase the remanence field of the permanent magnets. Permanent magnet-based insertion devices do not require a power supply to generate its magnetic field, but the magnetic field strength is limited by the remanence field of permanent magnets.
- **Low-temperature superconductor magnet-based insertion devices**: Several accelerator-based light sources are using low-temperature superconductor magnet-based various types of insertion devices including wavelength shifter [28–30], wigglers [31], superbends [32], and undulators [33, 34]. Low-temperature superconductor magnets outperform the high-field insertion devices such as wavelength shifters, wigglers, and superbends, and they are also comparable to the permanent magnet-based undulators. All the devices under operation use the NbTi for the low-temperature superconductor material. Recently, Nb_3Sn insertion device has been fabricated to further increase their performance and under preparation for the installation [35]. Most of the devices rely on liquid helium cooling which may increase the magnet operation cost.
- **High-temperature superconductor magnet-based insertion devices**: HTS magnets are yet to be operated in a real accelerator including accelerator-based light sources. Several studies fabricated HTS insertion devices to show their feasibility. Figure 1.7 shows selected recent studies on HTS insertion devices including: (a) 0.6 T, 4.2 K, magnetic period length(λ_u) = 16 mm, and engineering current density(J_e) = 2.1 kA/mm^2 by Argonne National Laboratory [23]; (b) 77 K, λ_u = 13 mm, and J_e=113 A/mm^2 by Chinese Academy of Sciences [24]; (c) 0.6 T, 4.2 K, λ_u = 13 mm, and J_e=2.3 kA/mm^2 by CERN and Karlsruher Institute of Technology

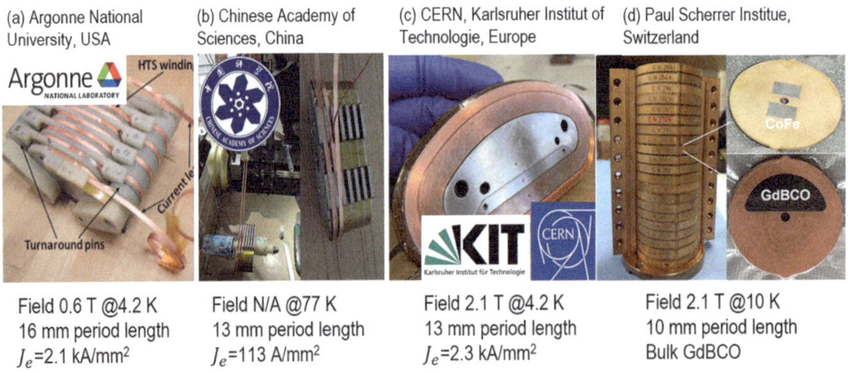

Fig. 1.7 Selected studies on HTS insertion devices by **a** Argonne National Lab [23]; **b** Chinese Academy of Sciences [24]; **c** CERN and Karlsruher Institute of Technology [25]; **d** Paul Scherrer Institute [26]

[25]; and (d) 2.1 T, 10 K, λ_u = 10 mm, and bulk-type undulator by Paul Scherrer Institute [26]. The developed HTS insertion devices are practically unavailable to install in the actual accelerator. For example, the studies in Fig. 1.7a–c fabricated and tested only half of the magnet structure. The bulk-type undulator in Fig. 1.7d generated the largest field of 2.1 T among others, however, magnetizing the bulk in real operation should be further investigated. Relatively low-field generation is mainly because most of the studies focus on undulator-type insertion devices with short magnetic period lengths. So far, high-field HTS insertion devices are yet to be demonstrated.

1.1.3 Basic Concept of Wavelength Shifter Magnet

A wavelength shifter magnet is a series of magnet with three magnetic poles to deflect electrons as shown in Fig. 1.5. The typical wavelength shifter magnet has one main pole and two side poles with opposite field directions to keep zero field integral. A strong field is generated at the main pole, therefore, the brightest light is generated at the main pole. Since the critical energy of the photon beam increases as the magnetic field increases, brighter photon flux can be obtained with a higher field. As the required photon energy increases, the brilliance of the photon that can be generated by the WLS magnet significantly increases.

Characteristics of photon beams are represented by spectral distributions of brightness as a function of the photon energy, where the brightness represents a number of photons per unit time, area, and solid angle at 0.1% bandwidth. Figure 1.8 compares spectral characteristics of the bending magnet and wavelength shifter magnet when applied to PLS-II. Estimated spectral characteristics of PLS-II are obtained

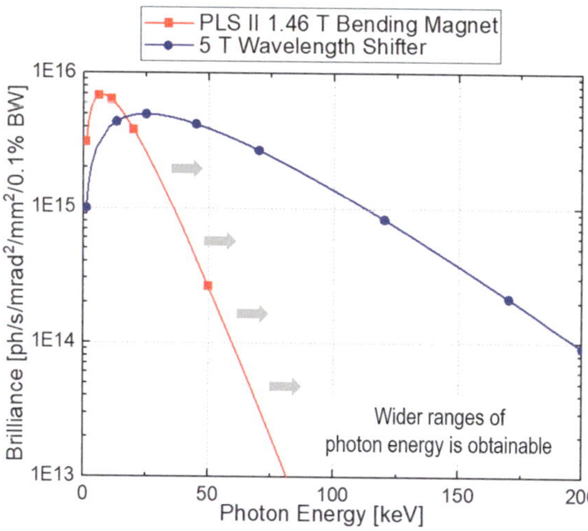

Fig. 1.8 Comparison of brilliance curve of radiated photon flux of 1.46 T normal conducting bending magnet and 5 T wavelength shifter magnet

from [36]. Wavelength shifter magnet may provide a brighter photon beam at higher photon energies. For example, over $>10^{15}$ of brilliance can be obtained up to 120 keV photon energy with 5 T wavelength shifter, while 30 keV for the conventional bending magnet.

1.2 Major Challenges of HTS Accelerator Magnets

Driving questions, key issues, and related challenges of the HTS accelerator magnets are as follows:

- **Driving question**: Can we apply high-field HTS magnets to a real accelerator?
- **Key issue 1**: Manufacturability of high-field HTS accelerator magnet.
- **Key issue 2**: Screening current-induced field effects of accelerator magnets.

The following subsections introduce technical challenges relating to these key issues in detail.

1.2.1 Mechanical Integrity of High-Field Accelerator Magnet

Superconducting magnets inevitably experience large Lorentz force due to their high current density and high magnetic field. Appropriate design of mechanical structure can guarantee the mechanical integrity of superconducting magnets. Figure 1.10 shows the measured critical current of REBCO conductors from different manufacturers depending on the applied stress [37]. The yield strength of REBCO conductor manufactured by SuNAM ranges from 600 to 800 MPa according to the previous studies [37–40].

Stress distributions in the coil are complex. The intensity and direction of Lorentz force, the shape of coil geometry, the existence of reinforcement structure, the stiffness of the material, and many other factors determine the stress distributions within the winding. In particular, non-circular coil geometry, often adopted for accelerator magnets to produce uniform magnetic along the long length of the beam trajectory, gives additional complexity to both stress modeling and reinforcement of the coil structure.

Figure 1.9 shows an example of a non-circular coil tested in the background field of 10 T [41]. The coil experienced critical current degradation after experiencing a large Lorentz force in the background field [41]. It is reported that the coil tended to deform into a circular section, as well the plastic deformation was observed after the disassembly of the coil. The degradation of coil characteristics is directly measured by the V–I transition curve tested in the liquid nitrogen in Fig. 1.9b. The coil obviously experienced I_c degradation according to the liquid nitrogen test results [42–44].

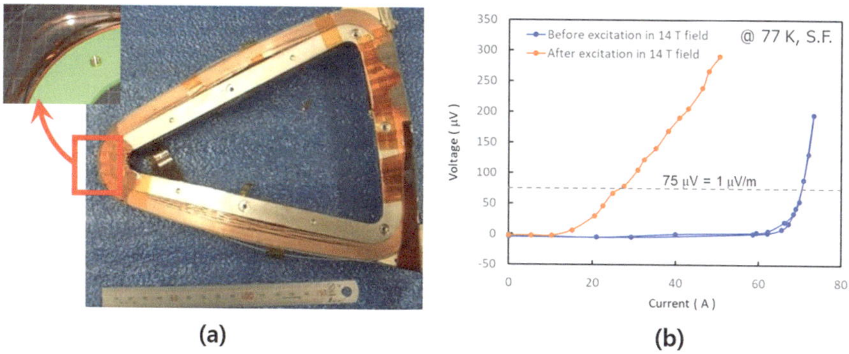

Fig. 1.9 **a** View of coil winding after powering test in 10 T background; **b** V–I characteristics before and after electromagnetic force test [41]

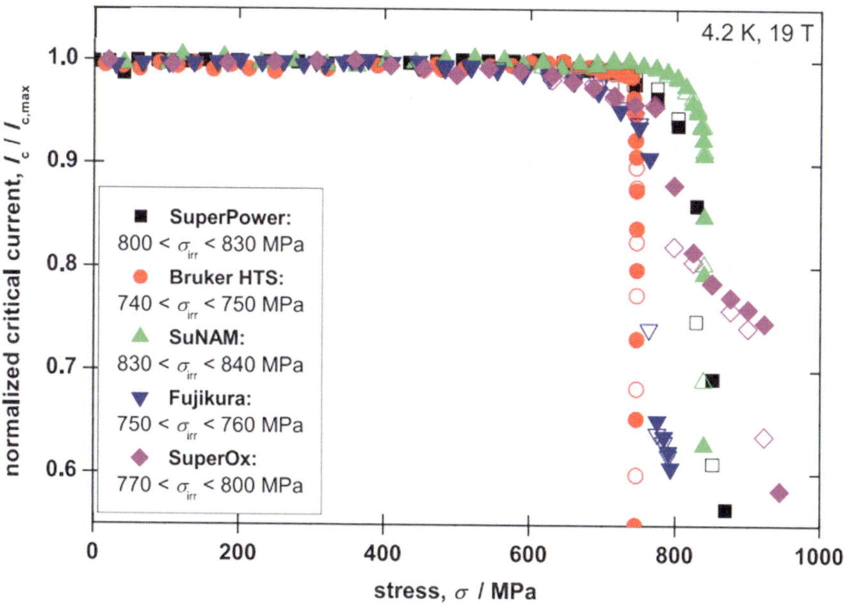

Fig. 1.10 Stress-dependent critical current of REBCO conductor measured at 4.2 K, 19 T [37]

1.2.2 High Cost of Liquid Helium Cooling for Magnet Operation

Maintaining cryogenic conditions is crucial for superconducting magnets to keep their performances because superconductivity occurs below a certain temperature. A practically usable cryogen for high-field superconducting magnets is liquid helium, mainly because low temperature is preferable for increasing the magnetic field performance.

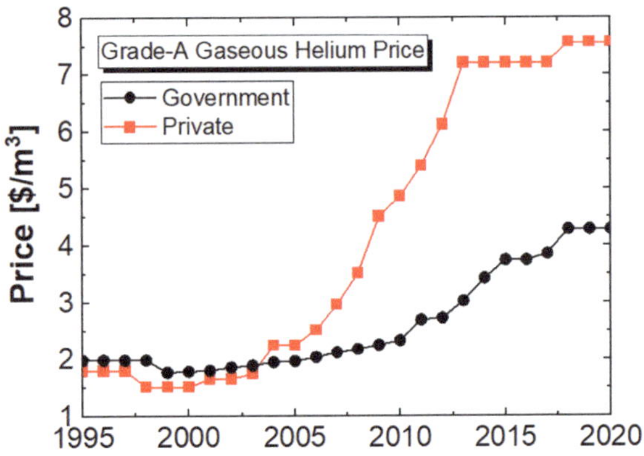

Fig. 1.11 Price of gaseous helium from 1995 to 2020 [45]

However, the high cost of liquid helium cooling systems is one of the challenges for superconducting magnets. Figure 1.11 shows the annual price of Grade-A gaseous helium released by the U.S. Geological Survey [45]. Since liquid helium is obtained by liquefying gaseous helium, the price of liquid helium also increases. The source of gas helium is natural gas, which is a limited resource. Since the increasing demand for limited helium resources in various industries, such as healthcare, electronics, and research, the price of helium is annually increasing.

1.2.3 Difficulties in Protection of HTS Magnets

The initiation of quench in superconducting magnets often starts in a small fraction of volume, called a hot spot. Stored energy in the magnet system may be released in the hot spot upon quench, leading to local temperature rise. Without adequate protection schemes of the superconducting magnet system, local temperature rise in the hot spot may result in permanent damage (e.g., burnout) to the superconducting magnet system. Detection of the quench should be preceded to activate the installed quench system.

Detection of the quench highly relies on the normal zone probation velocity (NZPv) of the superconducting magnet. However, early quench detection of HTS magnet is challenging due to the very low NZPv of HTS material. Typical ranges of NZPv of HTS are 1–10 mm/s, which is about 1,000 times slower than that of LTS [46]. The delay of quench detection due to the low NZPv makes it difficult to protect HTS magnets upon quench. Figure 1.12 shows an example of local burnout in the HTS magnet after the quench [47].

Fig. 1.12 Local burnout in HTS magnet after quench [47]

1.2.4 Effect of Screening Current-Induced Field in Beam Operation

Figure 1.13 shows a schematic to explain the principle of screening current and screening current-induced field (SCF) in a superconducting solenoid magnet with the coated conductor. The screening current is generated to cancel out the penetrating magnetic field generated by the transport current. Then, the field distributions in the target area are changed by the screening current, where both the field strength and uniformity may be changed. The magnitude of the screening current is proportional to the critical current and its effect is more dominant in coated-type conductors.

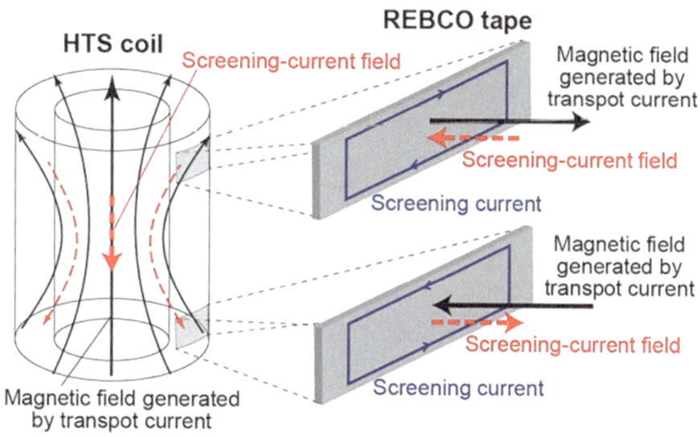

Fig. 1.13 Principle of screening current in coated HTS conductor [48]

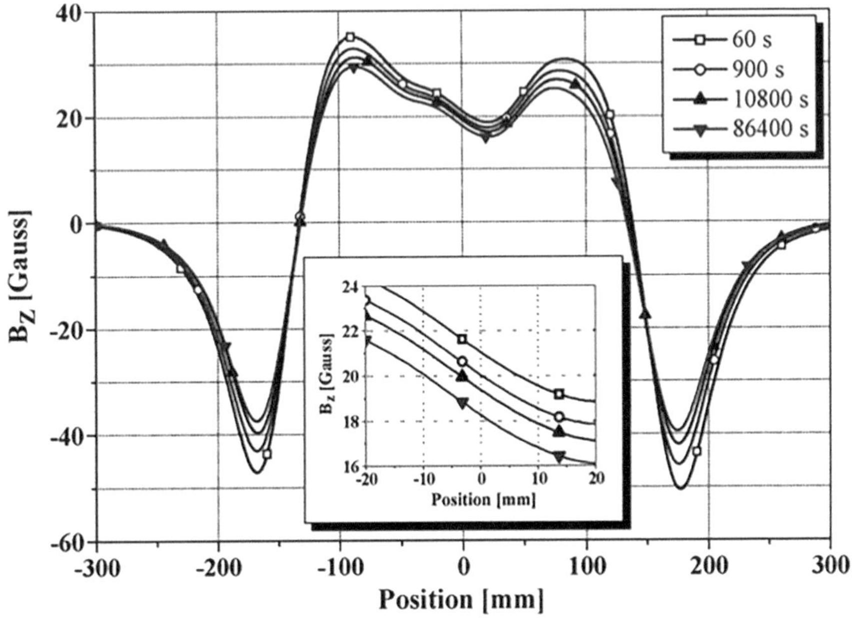

Fig. 1.14 Residual screening current-induced field measured in 100 MHz HTS insert [49]

One of the first studies to identify SCF in the magnet level may be 100 MHz HTS insert by MIT group. Figure 1.14 shows the measured residual spatial magnetic field distributions as well as the temporal change of field distributions after charging and discharging the magnet [49]. For this particular application, nuclear magnetic resonance, field uniformity at the target spherical volume is of particular importance. Otherwise, the detrimental effects of SCF may differ by the magnet shape and target applications. Therefore, various studies have been performed to estimate SCF in dedicated applications such as high-field research magnets, nuclear magnetic resonance magnets, magnetic resonance imaging magnets, dipole magnets, and many other industrial applications [50–56]. However, effects on SCF regarding the insertion devices are limited to simulation studies [57, 58]. In addition, none of the HTS magnets has yet to be operated in an actual accelerator.

1.3 Goal and Significance of The Study

1.3.1 Goal of the Study

Goal 1: Reinforcement and Preload Structure Dedicated for WLS Magnet

We aim to develop a mechanically robust WLS magnet system with a coil reinforcement structure as well as a preload structure. In this study, we cover the design, analysis, and manufacturing methods of such structures. The main purpose of the coil reinforcement structure and preload structure is to prevent overstress and maintain prestress against electromagnetic force and thermal contraction, respectively.

Goal 2: Liquid Helium Free Magnet System With Conduction Cooling Operation

Appropriate design and analysis of conduction cooling channels to minimize temperature gradient is one of the major goals of this study. Using the conduction cooling machine, electricity is only required to operate the cryocooler without any periodic replenishment of liquid helium. Therefore, the conduction cooling magnet system may be advantaged by its convenience in maintenance and operation. However, the major issue for the cryocooling system is a large temperature gradient within the magnet, which becomes more challenging with the increased size of the magnet. Since the only possible cooling source of the magnet is conduction, a detailed heat load analysis and thermal design are required for the conduction-cooled magnets.

Goal 3: Metal-Insulated Winding Method for Magnet Self-Protection

The idea to fabricate an HTS coil without turn-to-turn insulation came up in 2011, so-called the non-insulated HTS magnets [59]. Key features of the non-insulated HTS magnets are (1) self-protection mechanism by the current bypass against the hot spot; (2) mechanical robustness with the elimination of low stiffness insulation material; and (3) compactness with increased engineering current density. Owing to these advantages, several high-field magnets are constructed using the NI technology [60–77]. However, the charging delay of the magnet due to the leakage current becomes more severe with an increased size of the magnet, especially with a larger inductance. In this study, we adopted metal-insulated winding method to reduce the charging delay effect of non-insulated class HTS magnets.

Goal 4: Screening Current-Induced Field Analysis Model for Insertion Devices

Here, we develop nonlinear B-H characteristics considered screening current-induced field analysis model applicable to estimate three-dimensional field distributions for insertion devices. With the developed analysis model, the effect of SCF in beam operation is evaluated, such as the change of field integral. This thesis also discusses the operation scheme to reduce the screening current-induced field.

1.3.2 Significance of the Study

Suggestion of Design Methods of the Metal-Insulated Conduction Cooled HTS WLS Magnet

This research is the first trial to design and construct WLS magnet with HTS technology. Here we suggest the entire design process of the HTS WLS magnet including electromagnetic design, mechanical design, thermal analysis, and operation analysis. Accommodating the major challenges in mechanical integrity and cryogenic system configuration, this paper contributed to the following tasks in the design process:

- Propose design methods of WLS magnet with HTS technology.
- Provide a new structure and analysis model for straight section reinforcement of the racetrack magnet.

Suggestion and Demonstration of Core Manufacturing Technology for the First WLS Magnet with HTS Technology

This paper presents core manufacturing technologies to manufacture HTS accelerator magnets including winding, joint, structure, lead, cryogenics, and operation/evaluation. The developed WLS magnet is the first domestically produced superconducting insertion device in Korea. Additionally, this research provides test results of the record magnetic field of 5 T in HTS racetrack dipole magnets, where the largest field obtained from HTS accelerator magnet is 4.52 T so far [78]. Regarding the core manufacturing technologies, the following are the key contributions of this study:

- Define the manufacturing process of core component technologies.
- Develop the winding diameter estimation model in the winding process.
- Develop design and analysis methods of preload structure for HTS accelerator magnets.
- Demonstrate the conduction cooling test of the assembled magnet.

Proposal of the Analysis Model for Screening Current-Induced Field Effects in HTS Accelerator Magnets

Magnetic distribution along the particle trajectory is a particular concern for accelerator operation. The absence of modeling and experimental verification of nonlinear B-H curve considered screening current analysis model is main problem to be solved. We formulated H-formulation for a nonlinear B-H curve considered screening current analysis model implemented by the finite element method. After that, using the segmentation method in the previous literature [79], three-dimensional field distributions are obtained from the equivalent current loops. Finally, screening current-induced effects in beam operation and possible operation scheme is suggested.

- Develop a numerical method to calculate screening current-induced field distributions.
- Demonstrate the developed method with a constructed WLS magnet.
- Discuss SCF effects on accelerator-based light source operation.

1.4 Thesis Structure

Chapter 1 "Introduction" provides motivation for the study by describing the requirements of high-field magnets in accelerator-based light sources. Corresponding challenges for the implementation of high-field HTS magnets are also described. After that, the goal and significance of this study are prescribed against the mentioned challenges.

Chapter 2 "Theoretical Background" describes basic theories to understand the basic characteristics of superconductors, in particular REBCO conductors. To analyze the SCF effects in beam operation, the basics of beam physics based on linear optic theory are also presented.

Chapter 3 "Design and Analysis Methods of Iron-core MI HTS WLS Magnet" includes four different disciplines for superconducting magnet design: (1) electromagnetics, (2) mechanics, (3) cryogenics, and (4) operation analysis. Electromagnetic design process to maximize its performance with limited resources, mechanical analysis to reduce peak magnetic stress within the winding, thermal design to minimize temperature gradient, and lumped circuit analysis modeling for iron-dominated magnet are dealt in this section.

Chapter 4 "Core Manufacturing Technologies of HTS WLS Magnet" covers the entire process of manufacturing of HTS WLS magnet including: (1) winding, (2) joint, (3) structure, (4) lead, (5) cryogenics, and (6) performance evaluation. In the winding and structure description, newly adopted manufacturing concepts and modeling methods are shown. Otherwise, this section describes how the conventional HTS magnet manufacturing methods have been applied to this magnet.

Chapter 5 "Construction of HTS WLS Magnet" shows the fabrication and assembly process of the research target of this study. Individual liquid nitrogen tests for the manufactured HTS coils and assembly results are reported.

Chapter 6 "Conduction Cooling Test Results of HTS WLS Magnet" provides test results of the HTS WLS magnet to generate 5 T in conduction cooling operation. Measured temperature, voltage, and field during initial cooling and powering tests are presented. Screening current analysis results are compared by the measured field, and after that, SCF effects on beam operation are also discussed.

Chapter 7 "Conclusion" summarizes the conclusion of this thesis.

Chapter 2
Theoretical Background

2.1 Superconductivity

2.1.1 Type of Superconductor: Type I and Type II

A superconductor is a material that can conduct electricity with zero resistance. One of the unique characteristics of the superconductor is the Meissner effect that the superconductor does not allow a magnetic flux density to exist in its interior. Since the superconductor does not allow the magnetic flux to exist in its interior, the current flows at the surface of the superconductor considering Ampere's law. However, the magnetic flux cannot be abruptly decreased at the surface, but they flow in a very thin layer. This thickness is very small but they are important for characterizing the property of the superconductor. Figure 2.1 shows an external magnetic field exerted on the infinite slab of superconducting metal. At the boundary of the slab, magnetic flux density decays, and the penetration depth (λ) which characterized the decaying characteristic may be defined by

$$\int_0^{\infty} B(x)dx = \lambda B(0), \tag{2.1}$$

where $B(0)$ is the external magnetic field. The concept of coherent length (ξ) first came up by Pippard [80]. ξ characterizes the density of superelectrons at the surface between the normal and superconducting regions. The degree of order of superconducting phase is identified by the density of superconducting electrons. Superconductors can be divided into two types by the relation between penetration depth (λ) and coherence length (ξ). The classification of type I and type II superconductors may be formalized by

$$\begin{aligned} \sqrt{2}\xi &> \lambda \quad \text{(type I)} \\ \sqrt{2}\xi &< \lambda \quad \text{(type II)}. \end{aligned} \tag{2.2}$$

J. Park, *5 T High Temperature Superconductor 3 Pole Wavelength Shifter Magnet for Accelerator-based Light Sources*, Springer Theses,
https://doi.org/10.1007/978-981-96-8679-7_2

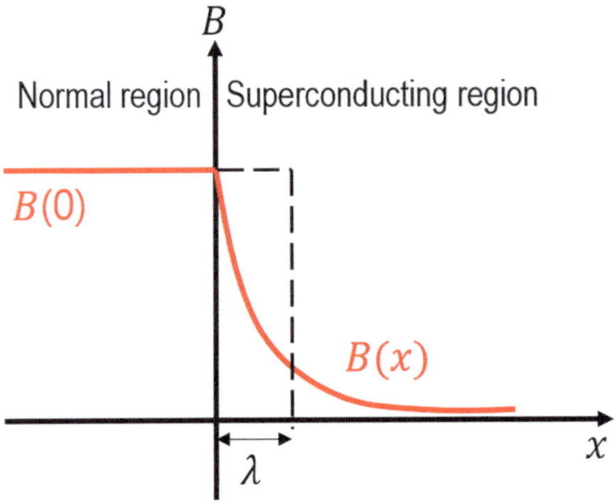

Fig. 2.1 Decaying of a magnetic field at the surface of a metal superconductor

Figure 2.2 shows the difference between type I and type II superconductors by the surface energy [81]. For the type I superconductor, the free energy increment by the magnetic contribution dominates the free energy decrement at the surface between normal and superconducting regions. The net free energy near the surface is positive for the type I superconductor, while negative for the type II superconductor. Due to the negative energy at the surface between the normal and superconducting regions

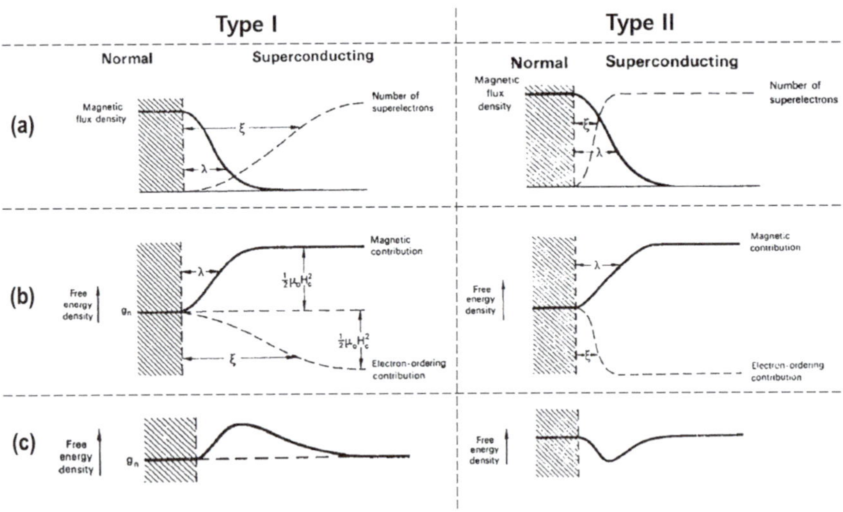

Fig. 2.2 Surface energy of superconductor: **a** penetration depth and coherence length; **b** contributions to free energy; **c** total free energy [81]

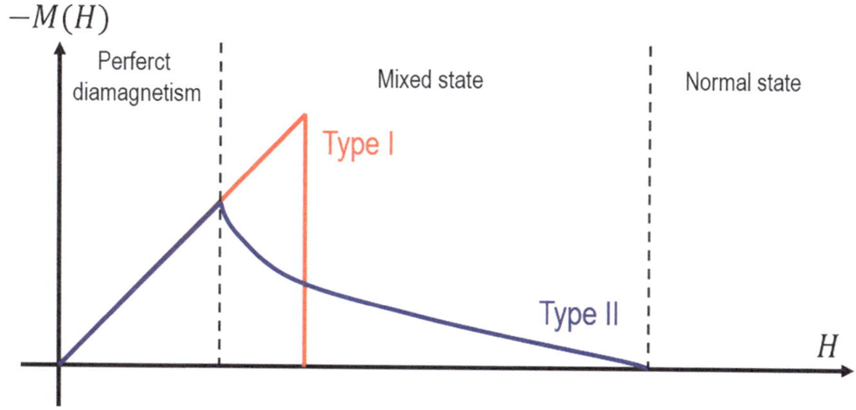

Fig. 2.3 Magnetization graph of type I and type II superconductor

of type II superconductors, it may be energetically favorable for type II superconductors to go into a mixed state. It forms magnetic field vortices when an external field is applied. A quantized flux called fluxoid penetrates into the superconductor when the external field increases beyond a certain field limit, and they do not show perfect diamagnetism. Figure 2.3 compares the magnetization curve of type I and type II superconductors. Type II superconductor shows perfect magnetism until H_{c1}, but goes to mixed state at the magnetic field beyond H_{c1}. In this region, type II superconductor still shows diamagnetism property, but not perfect magnetism.

Most type I superconductors are pure materials such as Hg and Al. However, due to the low critical field of around 0.1 T for type I superconductors, it is not possible to generate a high magnetic field with type I superconductors. Engineering-usable superconductors are mostly type II superconductors due to their better characteristics.

2.1.2 High-Temperature Superconductivity

Figure 2.4 shows the history of superconductor discovery [82]. The axis of the figure represents the discovered year and the critical temperature of the magnet, and the types of the superconductor are classified. The first superconductivity phenomenon was discovered in 1911 by Onnes. The superconductivity occurred at Hg at the temperature of 4.2 K. Among the thousands of superconducting materials, few of them are available for constructing magnets. Several requirements such as large critical current, large critical field, critical temperature higher than >4.2 K, mechanical strength, availability of manufacturing in wire type, and many other factors are very important for use in magnets. Typical examples of low-temperature superconductor (LTS) and high-temperature superconductor (HTS) materials used for the superconductor magnet are NbTi, Nb3Sn, Bi2212, Bi2223, REBCO, and MgB2.

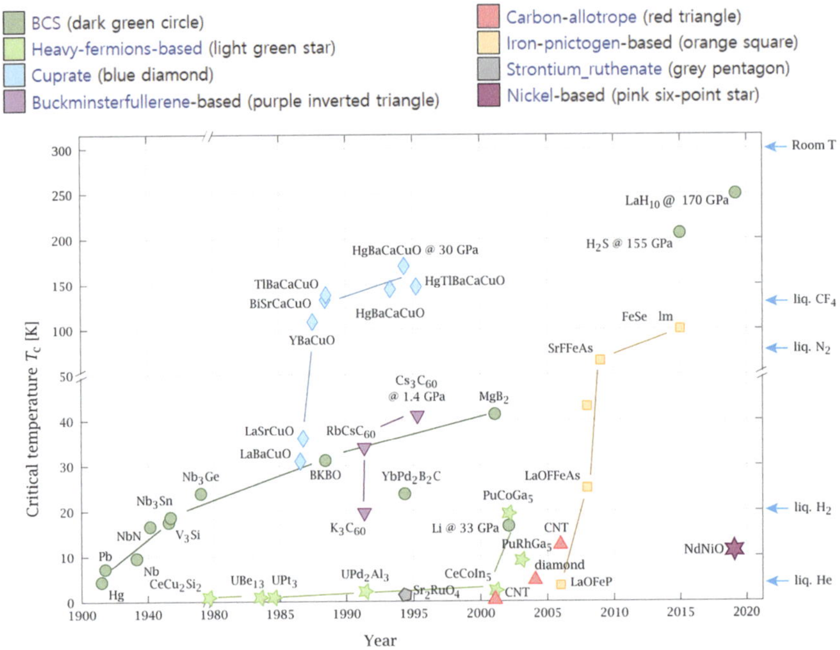

Fig. 2.4 Discovery of superconductor materials [82]

The term, HTS, is often used for superconductors having a critical temperature higher than the boiling temperature of liquid nitrogen, 77 K. This is mainly because the price of liquid nitrogen is way cheaper than the liquid helium which has been mostly used to cool down the superconductor magnets. However, there are no absolute criteria for classifying the LTS and HTS. MgB2 material is often referred to as HTS material, although the critical temperature of MgB2 is around 39 K.

2.2 Critical Current Estimation of Superconductor Magnet

2.2.1 Critical Surface of Superconductor

Superconductivity remains under certain conditions such as temperature, field, and current density. Figure 2.5 shows the critical surface for a typical type II superconductor. Superconductivity is maintained when the superconductor is operating under this surface. H_c, J_c, and T_c represent the critical field, critical current density, and critical temperature of the superconductor. In the engineering point of view, the temperature of the superconductor system should be kept very low, and cryogenic engineering

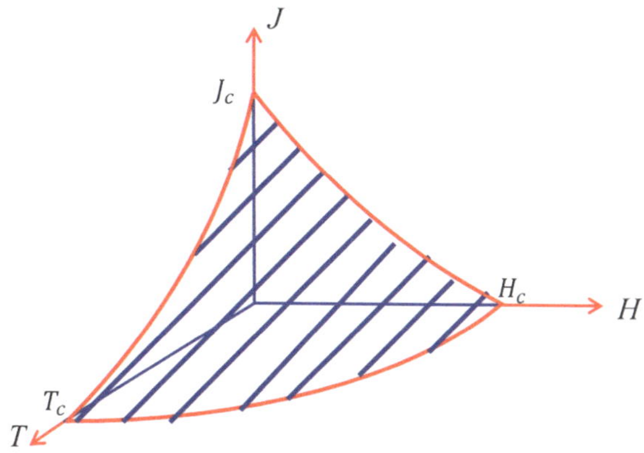

Fig. 2.5 Critical surface of superconductor

should be incorporated in such systems. In addition, the operating current of superconductor magnet is determined by both the J_c and H_c. To practically estimate the operating current of the magnet, the critical current density of the superconductor is measured at different field ranges.

2.2.2 *Anisotropic Characteristics of REBCO Critical Current*

Structure of REBCO Coated Conductor

Cuprate superconductors refer to the superconductor materials having a layer of copper oxides alternating with layers of other metal oxides. The mechanism of superconductivity in cuprate superconductors is not well understood, but they show high critical temperatures at atmospheric pressure compared to the other types of superconductors. Due to the layering structure of the cuprate superconductors, they have anisotropic field dependency on critical current characteristics.

The typical topology of REBCO conductor is shown in Fig. 2.6. The manufacturing process of REBCO conductor is similar to the semiconductor process, where the superconducting layer is deposited on the substrate and the buffer layer [83]. The thickness of REBCO layer is only in orders of 1 μm, while the total thickness of the conductor is mainly affected by the thickness of the substrate and stabilizer. The thickness of the substrate ranges around 30 to 100 μm by the manufacturer, and recently REBCO conductor with a thickness of 30 μm has also been reported [84]. Silver layer is deposited to cover the REBCO conductor by the sputtering method. After that, stabilizer material is electroplated on the conductor to protect the outer layer of the superconductor. Different kinds of materials such as copper, brass, and stainless steel are available [85], and thickness can be controlled depending on the demand of the user.

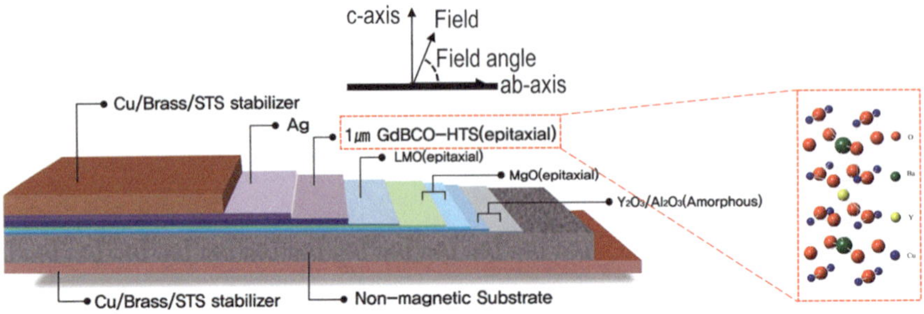

Fig. 2.6 Typical architecture of SuNAM REBCO conductor [83]

Measured Critical Current of REBCO

The critical current characteristics of superconducting wire vary by the manufacturer. To practically use the superconductors, their properties are measured at certain field and temperature conditions. In the case of REBCO conductors, field angle is an additional consideration since the critical current dramatically decreases with the higher c-axis field. Figure 2.7 shows the 20 K measured critical current of the REBCO

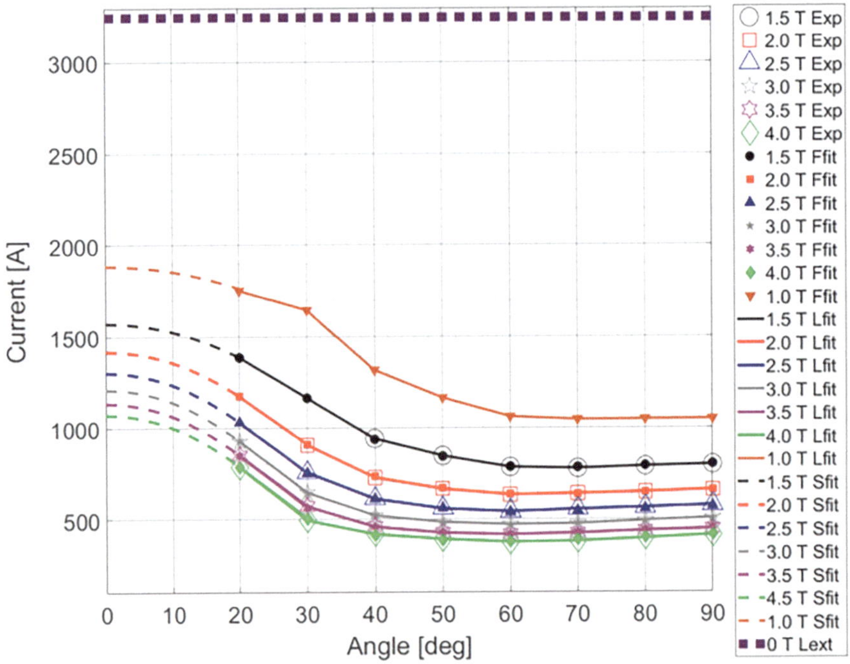

Fig. 2.7 Interpolated and extrapolated critical current from the measured critical current [50]

conductor manufactured by SuNAM [50]. The self-field critical current at 77 K of the measured sample is 600 A and the width of the sample is 12 mm. The critical current is measured at some selected field strength and angles. Using the curve fitting methods, $I_c(B, \theta)$ of 12 mm SuNAM REBCO tape at 20 K is obtained. The used fitting function and values are available in [50].

2.2.3 *Estimation of Critical Current: Load Line Method*

The critical current of a superconductor is limited by the external magnetic field. In the case of a superconductor magnet, as the operating current increases, the experienced magnetic field of the superconductor increases. As a result, the critical current of the superconductor decreases. Therefore, magnetic field analysis should be implemented to estimate the critical current of the magnet. One of the known approaches is to draw the load line analysis as shown in Fig. 2.8. The black solid line shows the relation between the magnetic field and the operating current of the magnet. At a selected operating current, the critical current is estimated by the magnetic field distributions within the winding. The critical current of the magnet is represented by the red line. If the magnet does not contain material with nonlinear magnetic properties, the load line analysis only requires magnetic field analysis once. The operating current at the intersection point is defined by the load line current (I_L). The operating current (I_{op}) is determined by considering the load line margin to I_L.

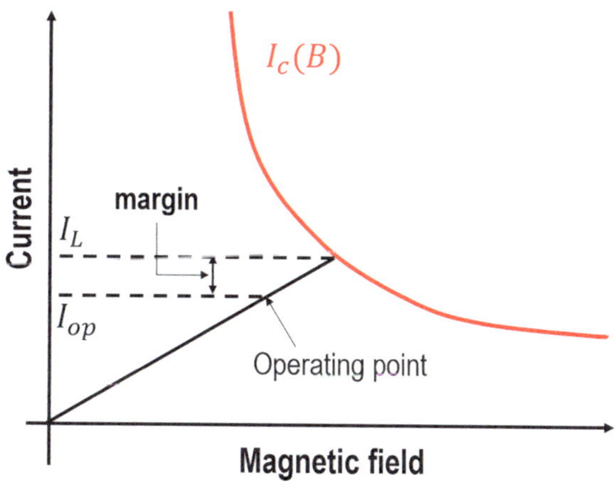

Fig. 2.8 Example of load line analysis

2.2.4 Multi-width Technique to Maximize Critical Current

The critical current of the superconductor magnet relies on the field distributions within the winding. Each winding pack experiences different magnetic field distributions, and as a result critical current differs spatially within the winding. An effective conductor grade may enable the magnet to be more compact or increase its magnetic field with a higher operating current. One of the suitable and effective approaches to conductor-grade pancake wound HTS magnets, multi-width technique has been suggested [86].

Most 2G HTS conductors are fabricated in the shape of tape, mostly slit from 12 mm width tape. Commercially, 2 mm to 12 mm width REBCO conductors are available. The key idea of the multi-width technique is to place different-width double pancake (DP) coils considering the field distribution within the winding. For example, the DP coils with the narrowest width are placed at the midplane of the magnet, where the *c*-axis magnetic field is the smallest. The width of DP gradually increases as further away from the center. The overall current density of DP coil assembly can be significantly increased by the MW technique. Figure 2.9 shows an example of MW configuration for 7 T MW NI REBCO magnet [87].

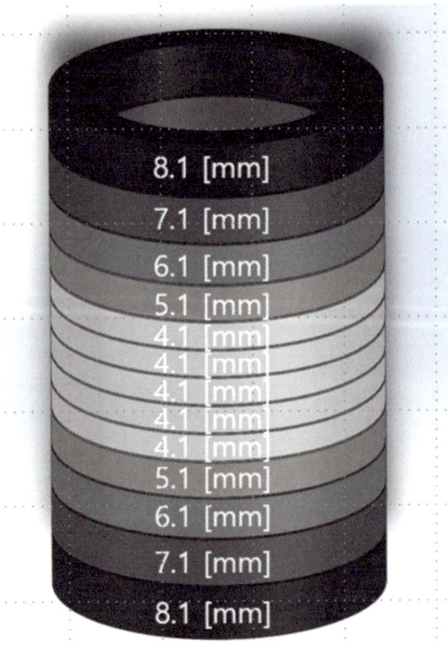

Fig. 2.9 Multi-width configuration of 7 T MW NI magnet

2.3 Basics of Beam Physics

2.3.1 Linear Equation of Particle Motions

The equation of particle motion in magnetic field region may be given by

$$\vec{F} = \frac{d\vec{p}}{dt} = m_0\gamma\frac{d}{dt}\vec{v} = q\vec{v}\times\vec{B}, \tag{2.3}$$

where $\vec{p}$ and $\vec{v}$ are particle momentum and velocity vector of the accelerated particle. Figure 2.10 defines the coordinate of the accelerator magnet. x-, y-, and z-coordinates represent transverse, gravitational, and longitudinal directions, respectively. The magnetic field region is illustrated by the red-colored box from z_0 to $z_0 + L$. For example, the directions of the ideal dipole field (B_y) and particle acceleration are y- and z-axes. The particle velocity and magnetic field vectors are represented by

$$\begin{aligned}\vec{B} &= B_y\hat{y}\\ \vec{v} &= v_x\hat{x} + v_y\hat{y} + v_z\hat{z}.\end{aligned} \tag{2.4}$$

By combining Eqs. (2.3) and (2.4), the equation of particle motion is represented by

$$\begin{aligned}\frac{d^2x}{dt^2} &= -\frac{q}{\gamma m_0}v_zB_y,\\ \frac{d^2y}{dt^2} &= 0,\\ \frac{d^2z}{dt^2} &= \frac{q}{\gamma m_0}v_xB_y.\end{aligned} \tag{2.5}$$

For the charged particles accelerated up to the velocity of lights, it can be assumed that $v_z >> v_x, v_y$. With this assumption, time derivative may be expressed by

$$\frac{d}{dt} = \frac{d}{dz}\frac{dz}{dt} \approx \frac{d}{dz}v_z. \tag{2.6}$$

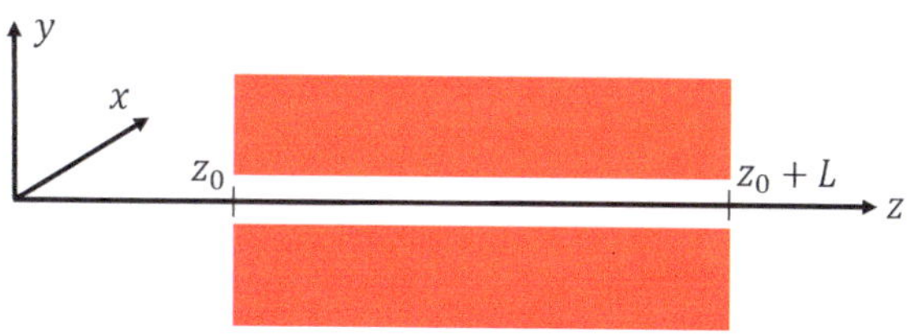

Fig. 2.10 Coordination definition of accelerator magnet

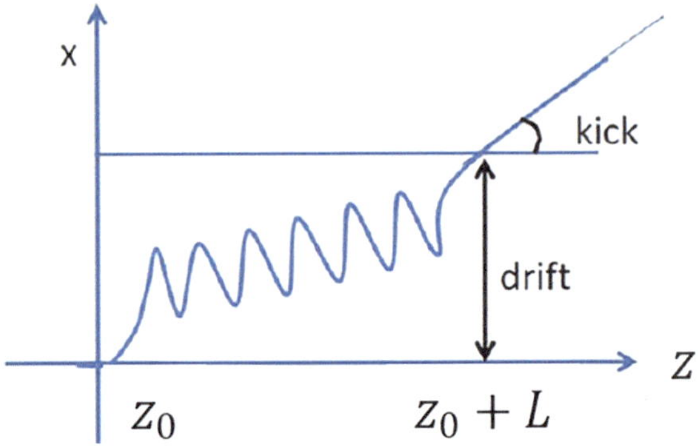

Fig. 2.11 Change of charged particle trajectory under magnetic field

Additionally, z-coordinate change of charged particles under a magnetic field can be neglected. Therefore, as the charged particle experiences a magnetic field, the x-trajectory of the particle changes as shown in Fig. 2.11. Both the deflection and location of the charged particle change after experiencing the magnetic field. The deflection(x') of the charged particle is defined by

$$x' = \frac{v_x}{v_z}. \tag{2.7}$$

The change of deflection (kick) and transverse coordinate (drift) of the charged particle can be obtained by integrating Eq. (2.5).

$$x'(z_0 + L) - x'(z_0) = \frac{v_x(z_0 + L) - v_x(z_0)}{\beta c} = \frac{q}{\gamma m_0 c}\int_{z_0}^{z_0+L} B_y(z')dz' \tag{2.8}$$

$$x(z_0 + L) - x(z_0) = -\frac{q}{\gamma m_0 c}\int_{\infty}^{z}\left[\int_{-\infty}^{z'} B_y(z'')dz''\right]dz'. \tag{2.9}$$

2.3.2 *Radiation from Relativistic Particles*

The first synchrotron radiation was observed by Floyd Habe in 1947 at the 70 MeV electron synchrotron. Relativistic charged particles emit very bright light when they experience acceleration. This electromagnetic radiation is called the synchrotron radiation. The total radiated power (P_s) by the relativistic particles can be formulated by

$$P_s = \frac{e^2 c}{6\pi\epsilon_0(m_0c^2)^2}\left[\left(\frac{dp}{d\tau}\right)^2 - \frac{1}{c^2}\left(\frac{dE}{d\tau}\right)^2\right], \tag{2.10}$$

where ϵ_0 and τ are the electric permittivity and proper time which is the time measured at the moving frame. Two types of acceleration, linear and circular, are possible for particles. The linear acceleration by the electric field also changes the momentum of charged particles in Eq. (2.10).

$$\frac{dE}{d\tau} = v\frac{dp}{d\tau}. \tag{2.11}$$

Therefore, the accelerated power by the linear acceleration due to the electric field is represented by

$$P_s = \frac{e^2 c}{6\pi\epsilon_0 (m_0 c^2)^2}\left(\frac{dE}{dx}\right)^2. \tag{2.12}$$

However, the radiated power by the linear acceleration is marginal considering the practical limit of the electric field. With a conventional value of 15 MeV/m for dE/dx, the radiated power is only 4×10^{-17} W.

Charged particle beams are circularly accelerated by the magnetic field. The power of synchrotron radiation by the circular acceleration is comparably larger than the one by the linear acceleration. The change of particle momentum with a bending radius of R is expressed by

$$\frac{dp}{dt} = \frac{1}{\gamma}\frac{dp}{d\tau} = p\frac{v}{R}. \tag{2.13}$$

Finally, the radiated power by the circular acceleration can be obtained by

$$P_s = \frac{e^2 c}{6\pi\epsilon_0}\frac{1}{(m_0 c^2)^4}\frac{\epsilon^4}{R^2}. \tag{2.14}$$

Increasing the beam energy can increase the radiated power, but the whole accelerator system should be upgraded. Another approach to increase the radiated power is by replacing the existing electromagnets with a higher field one. The radiated power by the circular acceleration is square proportional to the strength of the magnetic field since the bending radius is inversely proportional to the magnetic field strength.

2.3.3 *Spectral Distribution of Wiggler Radiation*

The radiated photon has wide ranges of energy ranges. The photon density spectrum emitted by the relativistic particle is formulated by Schwinger [88].

$$\frac{d\dot{N}}{dE/E} = \frac{P_0}{\omega_c \hbar} S_s\left(\frac{\omega}{\omega_c}\right)^2. \tag{2.15}$$

The left term in Eq. (2.15) is the photon spectral defined by the emitted number of photons per second and unit bandwidth. The ω_c and $\dot{N}$, and P_0 are critical frequency, photon density spectrum, and power radiated by the electron beam of current I_{beam}.

$$P_0 = \frac{e\gamma^4}{3\epsilon_0 R} I_{beam} \tag{2.16}$$

$$\omega_c = \frac{3c\gamma^3}{2R}. \tag{2.17}$$

The spectral function S_s is defined by the modified Bessel function ($K_{5/3}$).

$$S_s(\eta) = \frac{9\sqrt{3}}{8\pi}\xi \int_{\xi}^{\infty} K_{5/3}(\eta)d\eta. \tag{2.18}$$

The spectral function satisfies the following condition:

$$\int_0^1 S_s(\eta) = \frac{1}{2}. \tag{2.19}$$

Figure 2.12 shows an example of spectral photon density emitted by a 5 GeV electron beam circulating with a bending radius [3]. The physical meaning of ω_c can be explained by Eq. (2.19). The number of emitted photons below the critical energy is equal to the number of photons above the critical energy. The peak photon density occurs slightly below the critical energy.

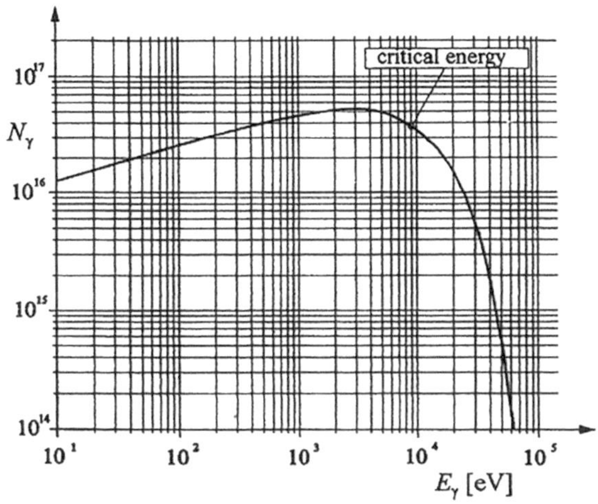

Fig. 2.12 Spectral photon density by synchrotron radiation from the bending radius of 12.2 m and 5 GeV energy [3]

The performance of emitted photon beams is generally represented by the brightness of the light (B_p). Brightness, also referred to as brilliance or spectral brilliance, is a total photon flux in a given six-dimensional space and unit bandwidth.

$$B_p = \frac{\dot{N}}{4\pi^2 \sigma_x \sigma_y \sigma_{x'} \sigma_{y'} d\epsilon/\epsilon}, \tag{2.20}$$

where σ_x, σ_y, $\sigma_{x'}$, and $\sigma_{y'}$ represent root mean square of beam size and solid angle of the photon beam.

Chapter 3
Design and Analysis Methods of Iron-Core MI HTS WLS Magnet

Key features of the designed WLS magnet are as follows: (1) use of metal insulation winding techniques for self-protection of the magnet as well as reducing charging delay; (2) multi-width configuration to maximize the performance of the magnet with limited conductor usage; (3) use of cold iron core to maximize magnetic field; and (4) liquid helium free conduction cooling operation to reduce operation cost. This chapter describes the design and analysis of the HTS WLS magnet considering the key features of the magnet.

Figure 3.1 shows considered physics for the design of the metal-insulated HTS magnet including electromagnetics, mechanics, cryogenics, and operation. The aim of the electromagnetic design is to maximize the center magnetic field with limited conductor usage and meet the field homogeneity requirement for accelerator operation. For the mechanical design, the new structure is suggested for straight section reinforcement of the racetrack magnet. Mechanical analysis methods suitable for the suggested structure are also presented considering the dry winding contacts of the coil. After that, conduction, radiation, and resistive heat loads are calculated, and thermal distributions of the WLS magnet are analyzed considering the conduction cooling operation. After that, the charging and discharging characteristics of the metal-insulated coil are analyzed by a lumped circuit model.

3.1 Electromagnetic Design of MI HTS WLS Magnet

3.1.1 Magnet Components to Generate Magnetic Field

3.1.1.1 Overall Geometry and Design Parameters

Figure 3.2 shows key electromagnetic components of WLS magnet including cover yoke, pole yoke, and HTS winding. The electron beam axis is defined by z axis,

J. Park, *5 T High Temperature Superconductor 3 Pole Wavelength Shifter Magnet for Accelerator-based Light Sources*, Springer Theses,
https://doi.org/10.1007/978-981-96-8679-7_3

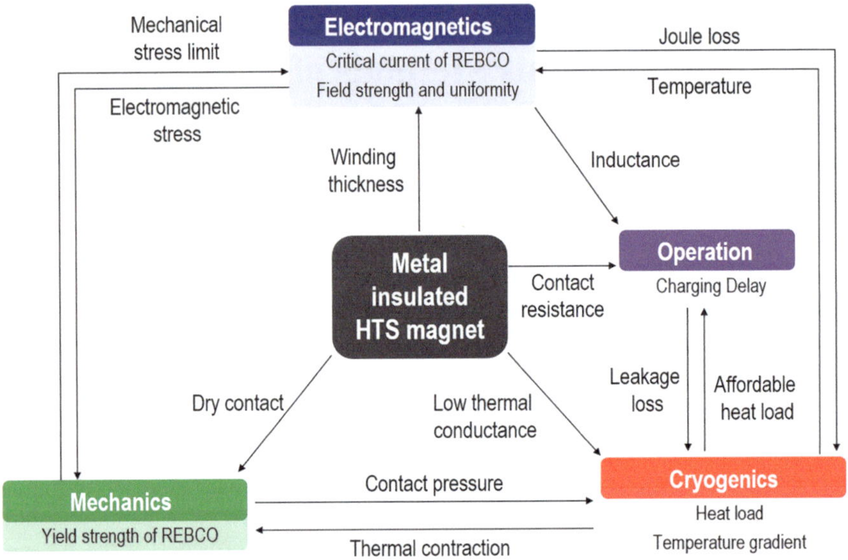

Fig. 3.1 Multi-physics considerations for metal-insulated HTS magnet design

Fig. 3.2 Schematic of wavelength shifter magnet which consists of HTS winding, cover yoke, and sidepole yoke

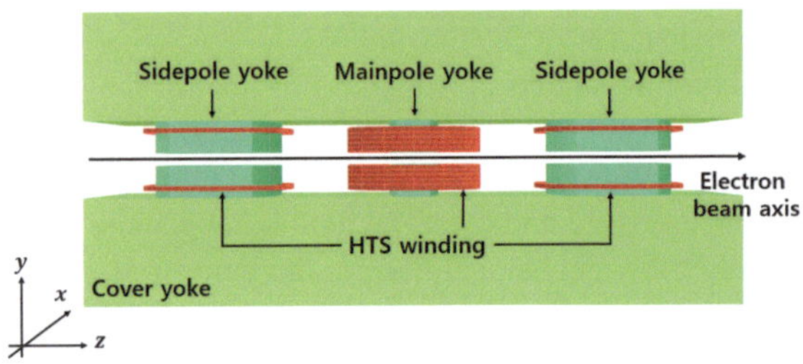

and HTS winding pack is piled up on y axis. Figure 3.3 shows detailed design parameters of WLS magnet. Design parameters include the number of winding turns for mainpole (N_M), and sidepole (N_S), straight section length of mainpole (l_M) and sidepole (l_s), number of mainpole stacks (N_{stack}), and width of conductor for each stack (w_i).

The size and shape of the cover yoke are rectangular shape width dimensions 300 mm (W_x) × 100 mm (W_y) × 670 mm (W_z). The magnetic gap (g), the distance between magnetic poles, is determined to be 12 mm considering operation in the PLS-II storage ring. To apply winding pressure at the straight section, the curvature of the straight section for both mainpole (r_{M_1}) and sidepole (r_{S_1}, r_{S_2}) is selected to be 700 mm.

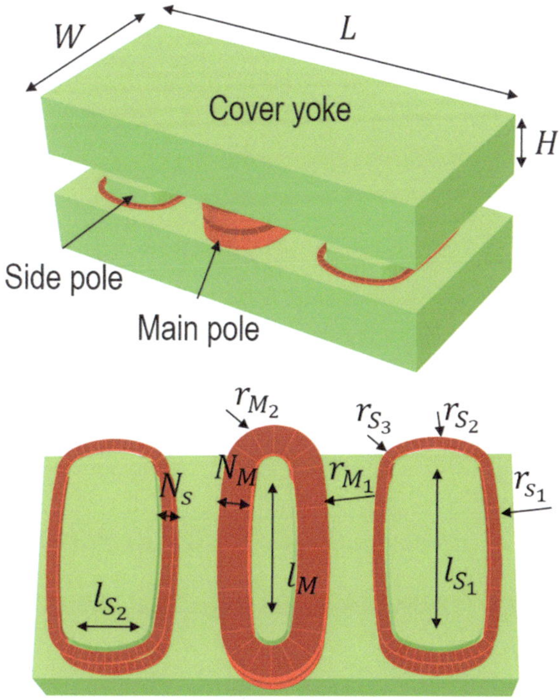

Fig. 3.3 Design parameter of WLS magnet

3.1.1.2 Material Choice for HTS Conductor and Iron Yoke

The specification of REBCO conductor varies by manufacturer, and the one manufactured by SuNAM is selected considering the supply chain. Standard 140 μm thickness SuNAM REBCO conductor copper-stabilized REBCO conductor and 60 μm thickness stainless steel tape are selected for co-winding. The REBCO conductor consists of stainless substrate of 105 μm (±4 μm) thickness, buffer layer of 5 μm, REBCO layer of 1 μm, and copper stabilizer of 20 μm.

The magnet consists of low-steel carbon cover yoke, permendur magnetic poles for both magnetic poles, and HTS winding pack. Permendur iron is used for pole yoke to concentrate magnetic field along the electron beam axis. The permendur has a high saturation field of 2.35 T, while 2.1 T for low-steel carbon. Figure 3.4 shows nonlinear B-H curve for both low-steel carbon and permendur used in the magnetic field simulation.

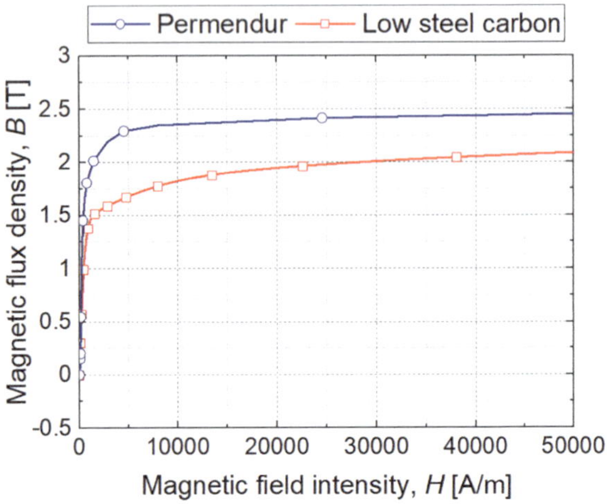

Fig. 3.4 Nonlinear B-H characteristics of low-steel carbon and permendur

3.1.2 Electromagnetic Design Requirements and Procedure

3.1.2.1 Design Requirements on Field Strength and Uniformity

Key electromagnetic design requirements are as follows: (1) main pole field of 6 T; (2) zero field integral along the beam axis; and (3) field integral uniformity of $<$ 100 G·cm at x=15 mm. Strength of main pole field is crucial for generation of high-flux photon beam at the high photon energy ranges. Since the insertion device is a transparent lattice device in the storage ring, the orbit of the electron beam should be maintained after experiencing the magnetic field. Therefore, zero field integral should be guaranteed in the design process. A condition for the field integral uniformity is also suggested to avoid nonlinear beam dynamics effect by the non-uniformity of the magnetic field distribution.

3.1.2.2 Flowchart of Electromagnetic Design Process

A flowchart of the electromagnetic design process is shown in Fig. 3.5. The electromagnetic design procedure consists of two parts: (1) an initial design to determine coil type with 2D FEM modeling; (2) a detailed design of coil shape optimization with 3D FEM modeling. Since the computational cost efficiency of 2D FEM modeling, wide ranges of various parameters can be investigated. However, field integral and its uniformity evaluation may not be feasible with 2D FEM. In the electromagnetic design part 1, type of main pole coil is determined, specifically N_M, N_{st}, and w_i are determined based on the 2D FEM modeling. After that, considering the field integral and its uniformity optimal values of l_{s_1}, l_{M_1}, and N_s were found.

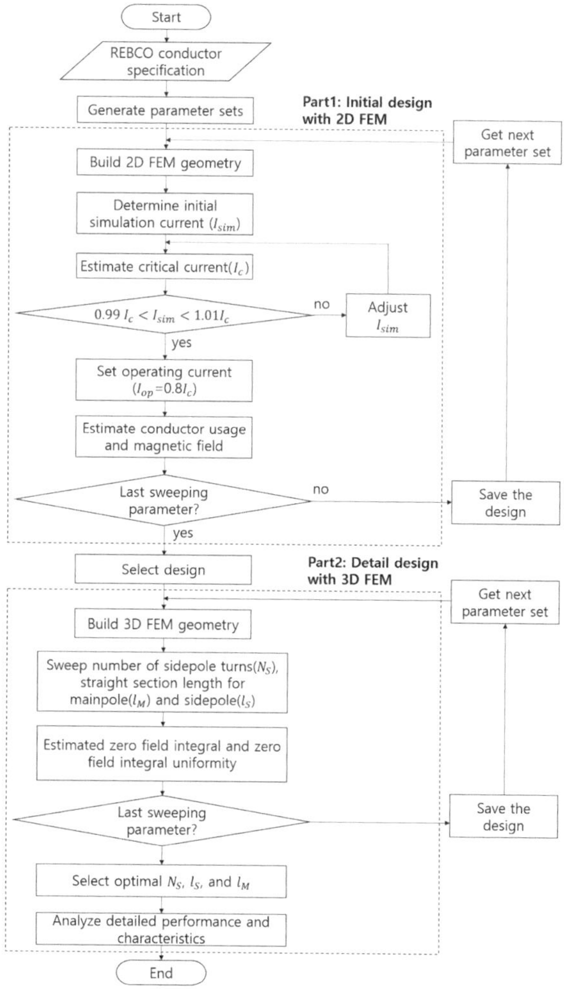

Fig. 3.5 Electromagnetic design flowchart of MI HTS WLS magnet

3.1.3 *Electromagnetic Design Results Based on Parameter Sweep Method*

3.1.3.1 Part 1: Initial Magnet Design with 2D FEM

Figure 3.6 represents geometry and boundary conditions used for 2D FEM simulation. Considering the symmetry of the magnet, only the quarter of the magnet cross section in y-z-plane is modeled. There is a 5 mm distance between the cover yoke and the uppermost main pole winding to place a cooling plate in between. Since the

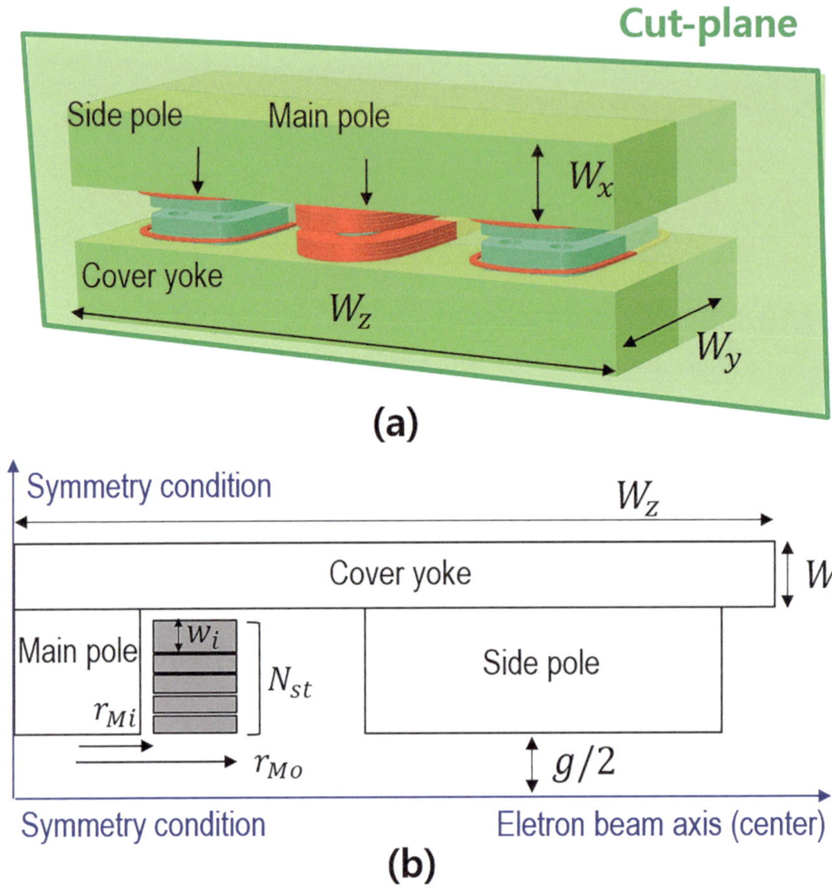

Fig. 3.6 **a** Cut plane and 3D view of magnet geometry, **b** Quarter of two-dimensional electromagnetic FEM analysis model considering magnet symmetry

purpose of the 2D FEM simulation is to determine parameters of main pole winding, main pole winding is only shown. The Neumann boundary condition at the symmetry axis is applied.

$$\nabla \times A = 0. \tag{3.1}$$

Parameter sweep method is used to determine the main pole winding configuration. Objective function ($F(x)$) and parameters are given by Eq. (3.2).

$$\begin{aligned} \textbf{Max.}\ & F(x) = (f_1(x), f_2(x)), \\ \textbf{subject to}\ & f_1(x) = B_M, \\ & f_2(x) = l_{REBCO}, \\ & x = (N_{st}, w_i, N_M). \end{aligned} \tag{3.2}$$

Table 3.1 Swept parameters and multi-width configuration

Parameters	Min.	Max.	Intervals
Number of main pole turns, N_M	100	300	25
Number of stacks, N_{st}	4	10	1
Width of conductor, w_i	4	5	1
Multi-width combinations (4 mm, 5 mm, 4 mm)			
(0, 4, 0); (0, 5, 0); (0, 4, 1); (0, 6, 0); (0, 5, 1); (1, 4, 1); (0, 4, 2); (0, 7, 0); (0, 6, 1)			
(0, 5, 2); (1, 5, 1); (1, 4, 2); (1, 3, 3); (0, 8, 0); (0, 7, 1); (0, 6, 2); (0, 5, 3); (1, 6, 1)			
(1, 6, 2); (1, 6, 3); (1, 5, 4); (2, 5, 2); (2, 4, 3); (0, 10, 0); (0, 9, 1); (0, 8, 2); (0, 7, 3)			
(0, 6, 4); (1, 8, 1); (1, 7, 2); (1, 6, 3); (2, 6, 2); (2, 5, 3); (2, 4, 4)			

Table 3.1 shows sets of swept parameters and multi-width combinations. A total of 442 sets of geometry with 34 sets of multi-width combinations are studied. REBCO conductors with thicker widths are placed at the edge of the winding pack since they experience a larger c-axis magnetic field.

Field- and angle-dependent critical current measured by Robinson Research Institute is used for determining the operating current of the design [89]. Until the difference between the simulation current and the estimated critical current is under 1%, simulation current is adjusted for a specific set of swept parameters. Considering the 20% load line margin of the magnet, operating current is determined for a specific parameter set.

Figure 3.7 shows parameter sweep results showing the relation between the main pole field and REBCO conductor usage (l_{REBCO}). Considering the electromagnetic design requirement (main pole field of 6 T), selected parameter set is (N_{stack}, N_M, H) = (5, 200, 100 mm), and total conductor usage is about 1.8 km. The selected design includes four stacks of 4 mm width winding pack and one for the 5 mm width. Therefore, it is decided to fabricate one single pancake coil for 5 mm main pole winding pack and four double pancake coils for the 4 mm one.

3.1.3.2 Part 2: Final Coil Configuration with 3D FEM

The conventional geometry of the racetrack coil is shown in Fig. 3.8a. A previous study reported that a slight curvature on the straight section may reduce the peak hoop stress of racetrack winding [90]. Another issue with the straight section winding is the difficulty of applying radial winding tension. Therefore, it is decided to modify the racetrack shape as shown in Fig. 3.8b, c, respectively, for mainpole and sidepole coils. Relation between the curvature of circular section (r_{M_2}, r_{S_2}, r_{S_3}), the curvature of straight section (r_{M_1},r_{S_1}), straight section length (l_M, l_{S_1}, l_{S_2}), straight section width (x), and angle (θ_M, θ_{S_1}, θ_{S_2}) is given by Eq. (3.3)

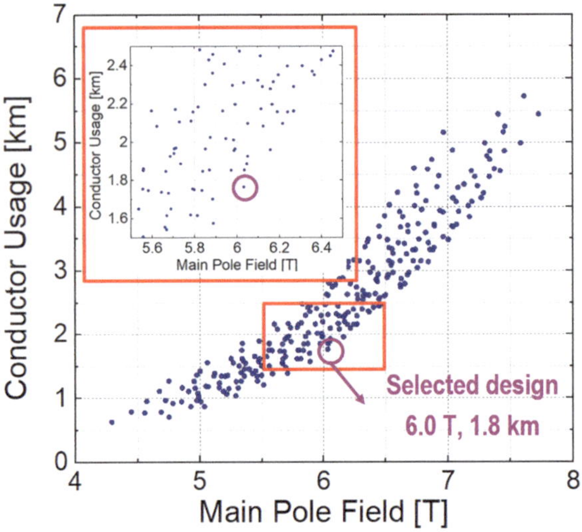

Fig. 3.7 Parameter sweep results with relation between main pole field and REBCO conductor usage

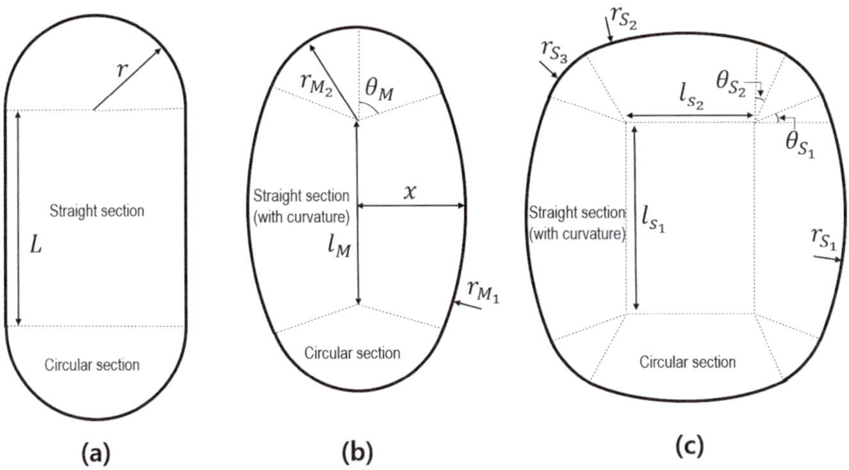

Fig. 3.8 **a** Conventional racetrack shape; **b** mainpole coil with slight curvature; **c** sidepole coil with slight curvature

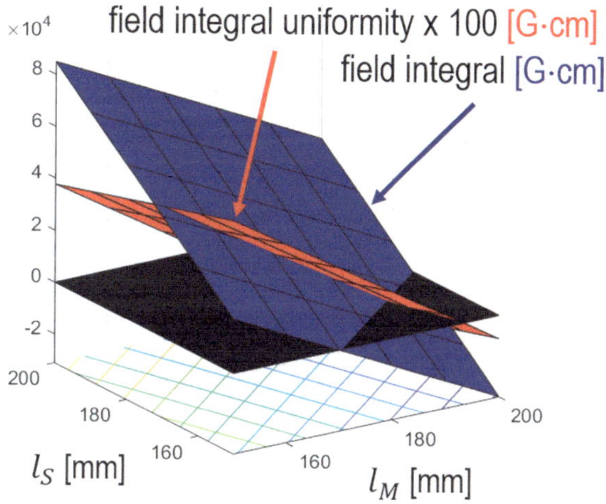

Fig. 3.9 Field integral and its uniformity depending on the straight section length of winding pack (l_S and l_M)

$$\begin{aligned}
\theta_M &= \sin^{-1}\left(\frac{l_M}{2(r_{M_1} - r_{M_2})}\right) \\
x &= R(1 - \sin\theta_M) = R\left(1 - \frac{l_M}{2(r_{M_1} - r_{M_2})}\right) \\
\theta_{S_1} &= \cos^{-1}\left(\frac{l_{S_1}}{2(r_{S_1} - r_{S_3})}\right) \\
\theta_{S_2} &= \cos^{-1}\left(\frac{l_{S_2}}{2(r_{S_2} - r_{S_3})}\right).
\end{aligned} \tag{3.3}$$

A slight curvature of 700 mm is given at the straight section for both side pole winding and main pole winding. r_{M_2} and r_{S_3} are selected to be 25 mm to keep bending strain under 0.2 %. The length of the straight section is different from the side pole and main pole. These values are determined to meet the field integral uniformity condition as well as the zero field integral.

Figure 3.9 shows field integral uniformity and field integral depending on the straight section length of the side pole and main pole coils. The blue and red surfaces represent field integral and field integral uniformity, respectively. The black surface represents a plane that meets the field uniformity condition: zero field integral and its uniformity. By finding a cross point of these three surfaces, l_M, l_S, and N_S are determined to be 188 mm, 166 mm, and 40.

3.1.3.3 Field Distributions of Final Design

Figure 3.10 shows the operating current-dependent main pole and side pole field of the WLS magnet. The slope of the magnetic field increase is large at the low operating current, but the slope dramatically decreases with the high operating current. This

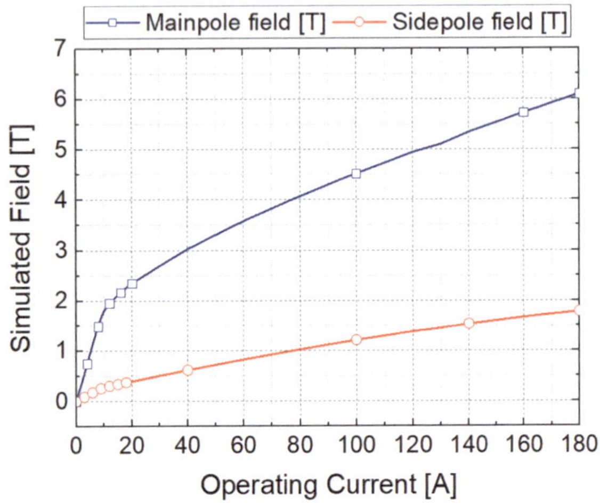

Fig. 3.10 Operating current-dependent center pole magnetic field

is mainly due to the saturation of iron magnetization. Final 3D field distributions and corresponding beam trajectories as well as the deflection angle are represented in Fig. 3.11. In the calculation of final field distributions, holes in the iron yoke for support installation are also considered. The main pole field, side pole field, maximum deflection angle, and maximum coordinate change are evaluated by 6 T, 1.8 T, 26 mrad, and 5 mm. For the calculation of deflection angle and beam trajectories, an electron beam energy of 3 GeV, which is the nominal operating energy of PLS-II, is assumed.

3.2 Mechanical Stress Analysis of Dry Wound Racetrack Coil

Mechanical integrity is crucial for the performance and reliability of the superconducting magnet system. Proper mechanical integrity ensures the magnet can withstand extreme operating conditions without deformation, damage, and failure. Irreversible stress and strain for 95% I_c retention of REBCO conductor are about 755 MPa and 0.68% according to the previous literature [37]. To keep the desired electromagnetic performance of REBCO magnets, local stress and strain values must be kept under these circumstances.

There are four different sources of mechanical stress in superconducting magnets: (1) winding stress during manufacturing of superconducting coil; (2) bending strain by the curvature of winding; (3) thermal stress during cooling down; and (4) electromagnetic stress by the Lorentz force. Especially, the magnitude of electromagnetic stress is significant for superconducting magnets due to their high field and high current density. For the racetrack superconducting coils, stress concentration at the

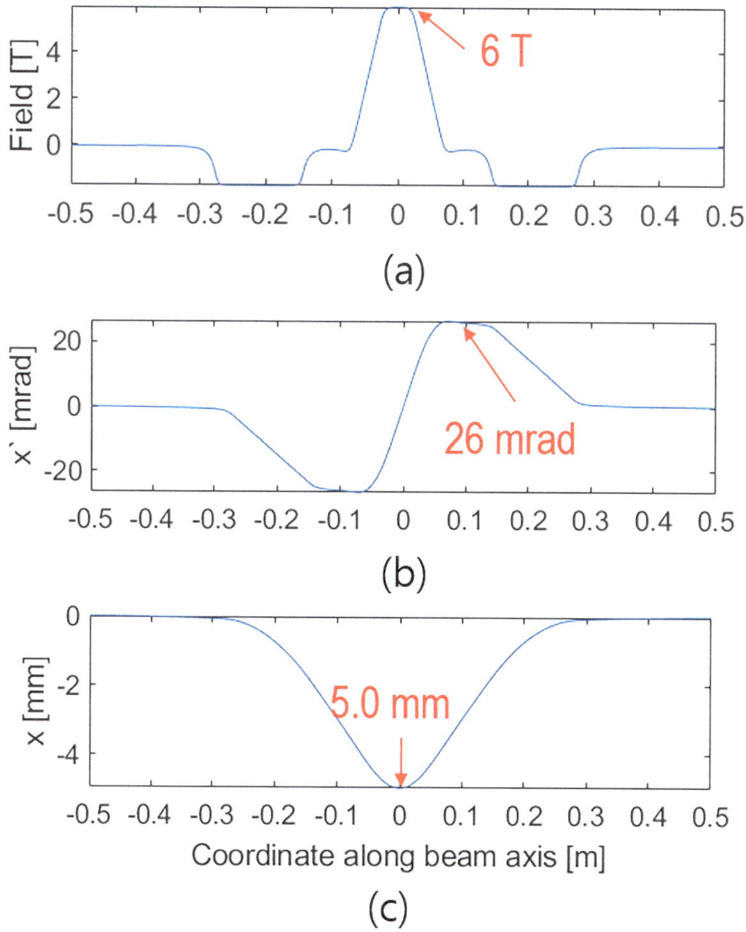

Fig. 3.11 **a** Field distributions along the beam axis; **b** deflection angle of electron beam; **c** beam trajectory

straight section of the winding is a particular concern. Figure 3.12 shows typical deformation of racetrack coils by Lorentz force [90]. Deformation in the racetrack coil occurs in the direction of making the coil to be more circular. Since the large expansion at the straight section, stress may be locally concentrated in the straight section winding.

In this chapter, the mechanical stress analysis results of the WLS magnet at 6 T operation are presented. FEM-based contact mechanics modeling to consider turn-to-turn contact of MI winding is implemented. A bolt pressurized straight section reinforcement structure is newly suggested to reduce the peak hoop stress at the straight section winding. As well as the suggested structure, a new mechanical analysis model is adopted to analyze the mechanical behavior of the HTS racetrack coil.

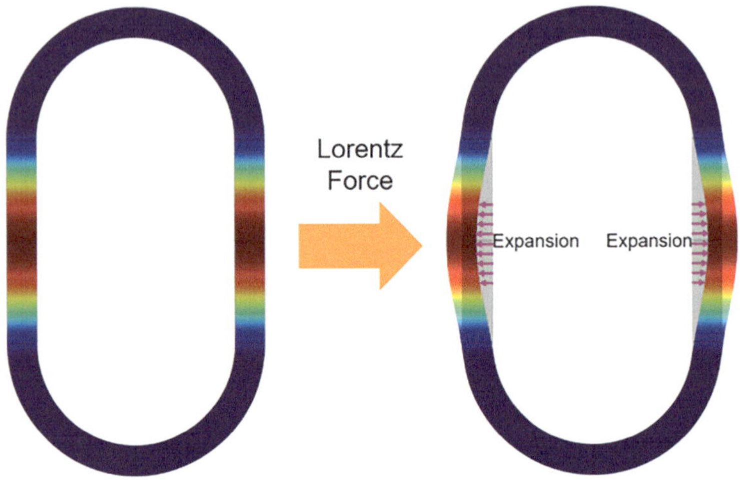

Fig. 3.12 Typical deformation of racetrack coil by Lorentz force [90]

3.2.1 Stress Mitigation by Straight Section Reinforcement

3.2.1.1 Major Challenges for Racetrack Reinforcement

Major challenges for the straight section reinforcement structure of racetrack coil are the effectiveness of the structure and space availability. If the structure is not properly attached to the winding, the structure may not facilitate its function. Also, additional space to install the reinforcement structure may reduce the compactness of the magnet. The key concept of the reinforcement structure is to pressurize the reinforcement structure by the bolt to keep the proper attachment. By electrically separating the winding pack and the reinforcement structure with an additional anodized aluminum, the reinforcement structure works as a cooling channel as well. Therefore, the space for cooling channel is only required for the reinforcement of the coil.

3.2.1.2 Suggested Reinforcement Structure for Conduction Cooled Racetrack Magnet

Figure 3.13 shows a CAD drawing of the reinforcement structure showing the front and rear views. The structure consists of a stud bolt, a buffer structure, and a reinforcement bar. The buffer structure is fabricated with aluminum and surface treatment of anodizing is applied after fabrication. This structure has a curved part to adhere closely to the winding, and the other side is flat to attach the main reinforcement bar. The reinforcement bar is made of oxygen-free high-purity copper (OFHC) which can also be utilized as a conduction cooling channel with its high thermal conductivity. The reinforcement bar has a screw thread on the side part to allow the stud bolt to

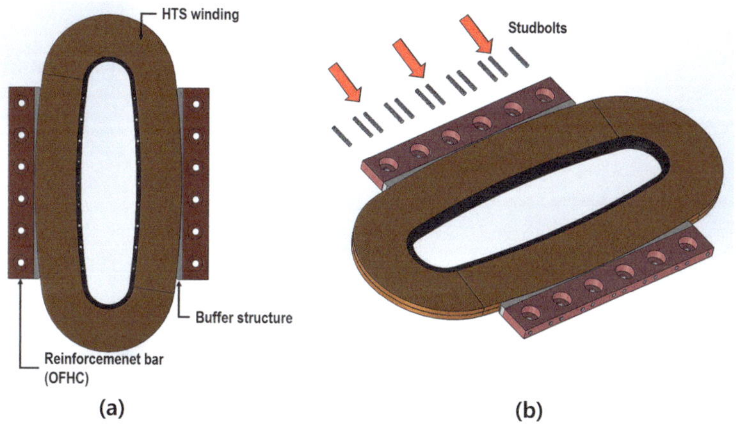

Fig. 3.13 3D CAD image of reinforcement structure: **a** top view; **b** rear view

penetrate the structure. The buffer structure has a grooved part at the flat surface to place the head of the stud bolt. After the assembly of the racetrack pancake coils and cooling channels, the stud bolt is screwed to apply pressure inside the coil. Finally, stud bolts allow the Lorentz force in the winding transfers to the reinforcement bar.

3.2.2 *FEM Modeling for Mechanical Stress Analysis*

3.2.2.1 Conventional Analysis Model for Dry Wound Racetrack Magnets

Based on the plane stress assumptions, a two-dimensional racetrack coil geometry model is constructed. Figure 3.14a shows FEM geometry and boundary conditions for mechanical analysis. The symmetry of the racetrack coil enabled only modeling quarter geometry of the winding by applying the roller boundary condition. The governing equation for mechanical analysis is given by Eq. (3.4).

$$\nabla\sigma + \vec{f}_v = 0, \tag{3.4}$$

where σ and $\vec{f}_v$ are stress tensor and Lorentz force density. Lorentz force density is calculated by the 3D electromagnetic FEM simulation results, and they are integrated along the winding volume. The roller boundary condition is given by Eq. (3.5), where displacement vector ($\vec{u}$) and normal vector ($\vec{n}$) are orthogonal.

$$\vec{u} \cdot \vec{n} = 0. \tag{3.5}$$

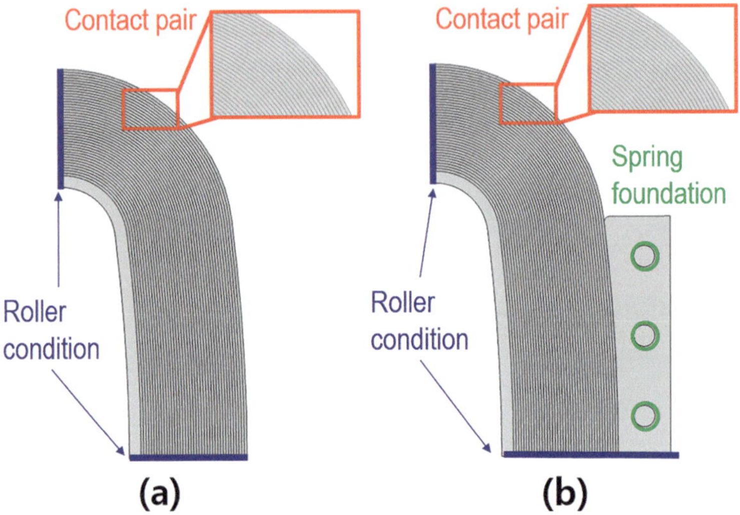

Fig. 3.14 FEM geometry for 2D FEM analysis for **a** without reinforcement structure; **b** with reinforcement structure

Table 3.2 Material properties for mechanical stress analysis

Material property	Aluminum	HTS winding	Copper
Young's modulus [GPa]	69	169 (E_r); 182 (E_θ)	80
Poisson ratio	0.3	0.34	0.32

To describe the turn-to-turn contact behavior of dry wound contacts, "contact pair conditions" are applied for every adjacent contact [53, 91, 92].

$$\begin{aligned} T_n &= 0\ (g_n > 0) \\ &= -p_n g_n\ (g_n <= 0), \end{aligned} \tag{3.6}$$

where p_n, g_n, and T_n are penalty factor, gap between the surface, and contact pressure. For the ease of computational cost, four turns are regarded as one engineering turn [93].

Table 3.2 summarizes material properties used for the simulation.

3.2.2.2 Modified Mechanical Analysis Model: Spring Foundation Condition

The reinforcement bar is fixed by a bolt which is screwed on the cover iron yoke. Figure 3.14 shows holes for placing the bolts. Lorentz force with radially outer

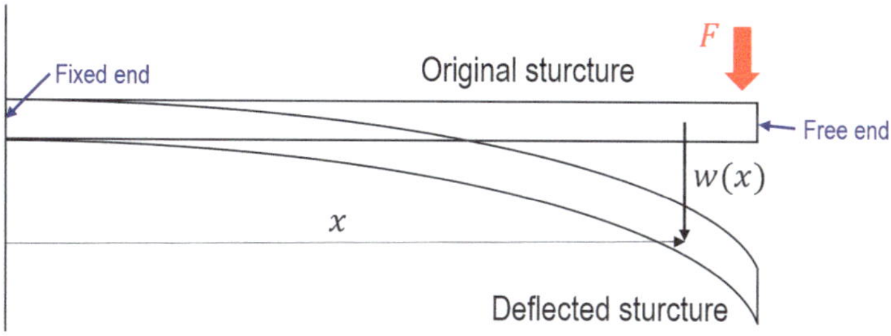

Fig. 3.15 Deflection of structure with the applied force

direction may also deform the location of bolts. Considering deformation behavior of long bolts, an additional boundary condition called spring foundation condition is given as shown in Fig. 3.14b. The spring foundation condition is represented in Eq. (3.7).

$$\vec{F_A} = -k_A(\vec{u} - \vec{u_0}), \tag{3.7}$$

where F_A, k_A, and $\vec{u_0}$ are reacting force, spring constant, and initial displacement.

The deformation behavior of bolts may be represented by long structural elements with one fixed end as shown in Fig. 3.15. The governing equation of beam deflection can be described by the Euler-Bernoulli equation [94].

$$\frac{d^2}{dx^2}\left(E_{iso} I_m \frac{d^2 w}{dx^2}\right) = F, \tag{3.8}$$

where w, x, E_{iso}, I_m, and F are transverse position, Young's modulus of isotropic material, mass, area moment of inertia, and distributed load. The boundary conditions for the fixed end and free end are represented by Eq. (3.9).

$$w|_{x=0} = 0, \quad \left.\frac{dw}{dx}\right|_{x=0} = 0, \quad \left.\frac{dw^2}{d^2x}\right|_{x=L} = 0, \quad \left.\frac{dw^3}{d^3x}\right|_{x=L} = 0. \tag{3.9}$$

When the load is distributed at the beam end, deflection of the homogeneous cantilever beam, a structural element that extends horizontally with fixed one end, can be obtained by Eq. (3.10).

$$w(x) = \frac{Fx^2}{6E_{iso}I_m}(3L - x). \tag{3.10}$$

The area moment of inertia for a homogeneous cylinder is calculated by

$$I_m = \int \rho^2 dA = \int_0^r 2\pi\rho^3 d\rho = 0.5\pi r^4. \tag{3.11}$$

Finally, the spring constant of the bolt with length of L and radius of r is obtained by dividing deflection over the applied force.

$$k_A = \frac{F}{w(L)} = \frac{3\pi E_{iso} r^4}{2L^3}. \tag{3.12}$$

The spring constant of the M6 stainless steel bolt with a 6 mm diameter and 20 mm length is 10 MN/m.

3.2.3 *Mechanical Stress Analysis Results with and without Reinforcement Structure*

Figure 3.16a–c shows deformation, calculated hoop stress, and radial stress of the main pole coil without reinforcement. To visualize the pattern of deformation, the deformed shape of the winding is also shown in Fig. 3.16a. The racetrack is deformed in the direction of becoming a circle. Without the straight section reinforcement

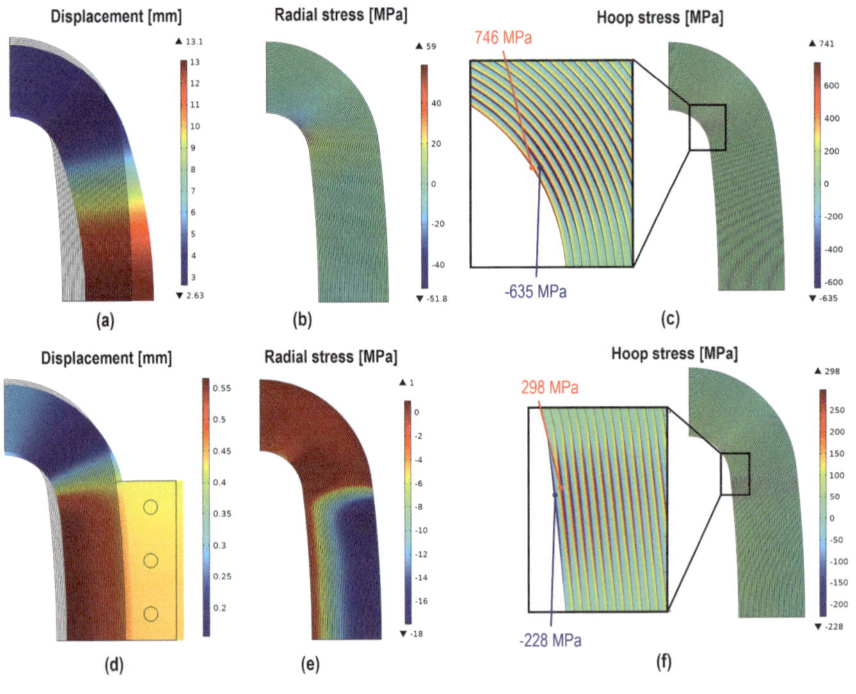

Fig. 3.16 Mechanical stress analysis results of main pole coil without reinforcement: **a** displacement; **b** radial stress; and **c** hoop stress. With reinforcement: **d** displacement; **e** radial stress; and **f** hoop stress

structure, hoop stress is estimated to be 746 MPa which exceeds the I_c retention stress of REBCO conductor. The peak hoop stress occurs at the innermost turn near the point where curvature changes. The calculated deformation is about 13.1 mm which may significantly change the field distributions. It is concluded that the coil cannot withstand electromagnetic stress without the reinforcement.

Therefore, mechanical reinforcement of the main pole coil is inevitable and it is decided to adopt the reinforcement structure as shown in Fig. 3.13. The size of the bolt used for fixing the reinforcement bar is an important parameter that is related to both the thermal and mechanical properties of the magnet. As the size of the bolt increases the stiffness of the structure increases and the structure contributes more to reduce the peak hoop stress. However, a larger through hole is required to install the bolt and it increases the thermal resistance between the stacked magnets. We investigated the relation between the size of bolts and peak hoop stress as shown in Fig. 3.17. The bending stress (σ_{bend}) can be calculated by

$$\sigma_{bend} = E_{\theta} d / 2D, \tag{3.13}$$

where E, d, and D are Young's modulus of the material, the thickness of the bending material, and the bending diameter. Considering the substrate thickness of 110 μ m, bending stress is calculated to be 200 MPa. To limit the maximum stress level of the magnet under 50 % of the irreversible stress (755 MPa), the diameter of the bolt is decided to be 6 mm. Finally, Fig. 3.16d–f shows the deformation, calculated hoop stress, and radial stress of the main pole coil with and without the reinforcement. Ten times magnified deformation is shown in Fig. 3.16d, and stress distributions within the winding are only plotted for Fig. 3.16e–f.

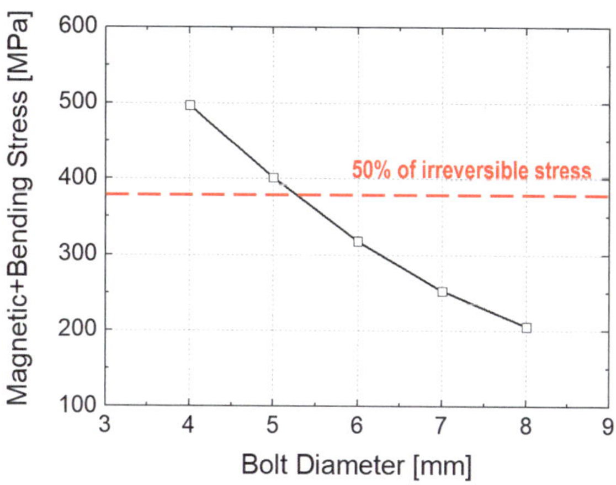

Fig. 3.17 Calculated maximum hoop stress depending on the bolt diameter

3.3 Thermal Analysis of Conduction Cooling System

3.3.1 Cryogenic System Configuration

Temperature is a crucial physical parameter in maintaining superconductivity. To keep the magnet at a desired temperature, the cryogenic system must extract all the heat sources of the magnet. Figure 3.18 shows conduction cooling system to test the WLS magnet. At the bottom of the conduction cooling system, conduction cooler model of Sumitomo RDK408 is installed. Figure 3.19 shows the cooling capacity of the given cryocooler [95]. Second stage heat capacity is 1.0 W at 4.2 K, while first stage heat capacity is 50 W at 43 K.

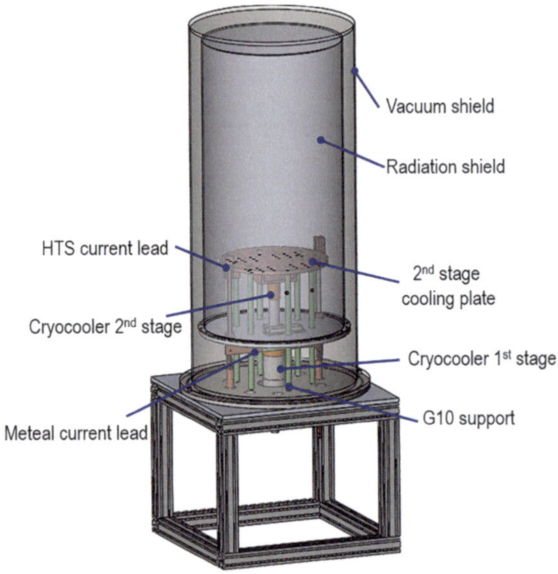

Fig. 3.18 In-house cryostat for conduction cooling test

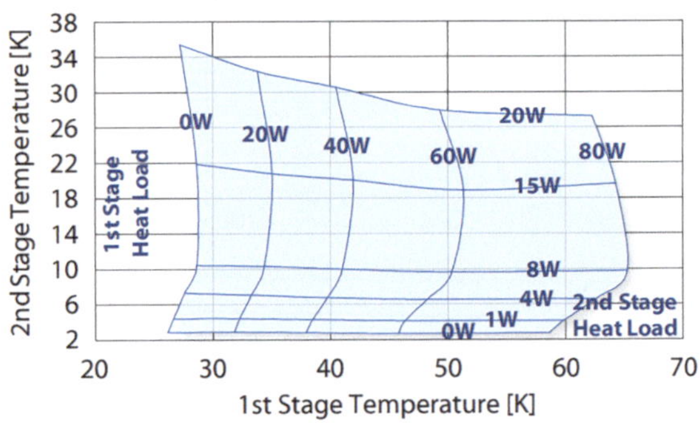

Fig. 3.19 Cooling capacity map of RDK408 [95]

The cryogenic system has a thermal shield to minimize heat loads around the magnet. The first stage, whose temperature is around 40 K, the radiation shield and a pair of two metal current leads are thermally connected. The second stage, where the magnet is installed, is thermally connected to the cryocooler second stage by the OFHC braid. The magnet is installed vertically above the main cooling plate. A set of HTS current leads are connected between the first and second stages to transport current to the magnet.

3.3.2 Heat Load Analysis Results

Table 3.3 summarized total estimated heat sources. Details on how to calculate each heat load are explained as follows.

- Conduction: Heat is intruded through conduction by the cryostat supports. The test facility has six GFRP rods to support the main cooling plate connected from room temperature and the second stage cooling plate. Heat equation for the conduction is given by

$$q_c = -k\nabla T, \tag{3.14}$$

  where T and Q are temperatures and rate of flow of heat energy per unit volume. Figure 3.20 shows the FEM modeling and calculated temperature distributions of these two structures. Boundary conditions of fixed temperature are given at each end of the structure. Finally, the heat load is calculated to be 0.5 W.
- Convection: Convection heat transfer (q_g) by the residual gases in very low pressure ranges is proportional to the pressure (P_g) level.

$$q_g = \eta_g P_g (T_w - T_c), \tag{3.15}$$

  where η_g is a constant of proportionality, and T_w and T_c are temperatures of warm end and cold end. For residual H_2 gas, η_g is 1.32 W/m^2·Pa·K. With a combination

Table 3.3 Calculated heat loads

Heat sources		Second stage	First stage
Conduction	G10 support	0.5	
	HTS lead support	0.34	
	Metal current lead		26.4
Resistive	Joint	$I_{op}^2 \times 4.5\,\mu\,\Omega$	
	Metal current lead		$I_{op}^2 \times 83\,\mu\,\Omega$
	lead	$I_{op}^2 \times 2\mu\,\Omega$	
Radiation		0.08	15.5
Total		$0.92 + I_{op}^2 \times 4.5\mu\,\Omega$	$41.9 + I_{op}^2 \times 83\mu\,\Omega$

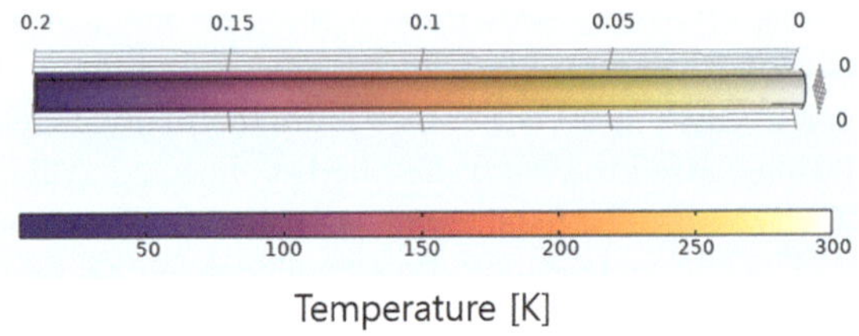

Fig. 3.20 Temperature distributions of support calculated by FEM

use of a rotary pump and turbo pump, pressure in the cryogenic system may be kept under 10^{-6} torr. Considering the total surface area of the system, 0.39 m^2, the estimated heat load by the convection is under 19.3 mW which is negligible.

- Radiation: To reduce the loss from the room temperature radiation, a radiation shield consisting of 25 polished multi-layer insulation (MLI) is installed in the first stage. Equation (3.16) shows the loss from the radiation (q_r).

$$q_r = \epsilon_r (T_w^4 - T_c^4)/N_i, \tag{3.16}$$

where N_i is number of insulation. Measured radiation heat of 40 mW/m^2K (40 MLI) and 2500 W/m^2K (60 MLI), respectively, for 40 K to 4 K and 300 K to 40 K radiation [46]. Total radiation loss is estimated to be 15.5 W and 0.08 W for first and second stages.

- Resistive loss: Resistive joint in between the pancake coils, inner current lead, a connection between the side pole and main pole, and a connection between inner current lead and HTS current may result in resistive loss for the second stage. The magnet requires nine resistive joints, and the value of the joint is assumed to be 500 nΩ per each of them.
- Current lead: Current lead electrically connects a power supply at room temperature and a magnet at cryogenic temperature. At the same time, they are mechanically connected and conduction loss happens. As the length/area ratio of metal current lead increases, resistive loss increases while the conduction heat intrusion decreases. Therefore, there is a balance in the optimal configuration of the current lead. Figure 3.21 shows temperature distribution analysis results for the current lead with fixed end temperatures when the operating current is 400 A. As a function of transport current (I_{op}), the heat intrusion in the first stage is calculated to be 26.4+ $(83\mu\ \Omega) I_{op}^2$.

3.3.3 *Thermal Analysis Result Using FEM Model*

Figure 3.22 shows FEM modeling of magnet structure thermally connected to the second stage. Heat sources described in the previous section are given as inputs. Thermal contact resistance is given as boundary conditions for every thermal contact.

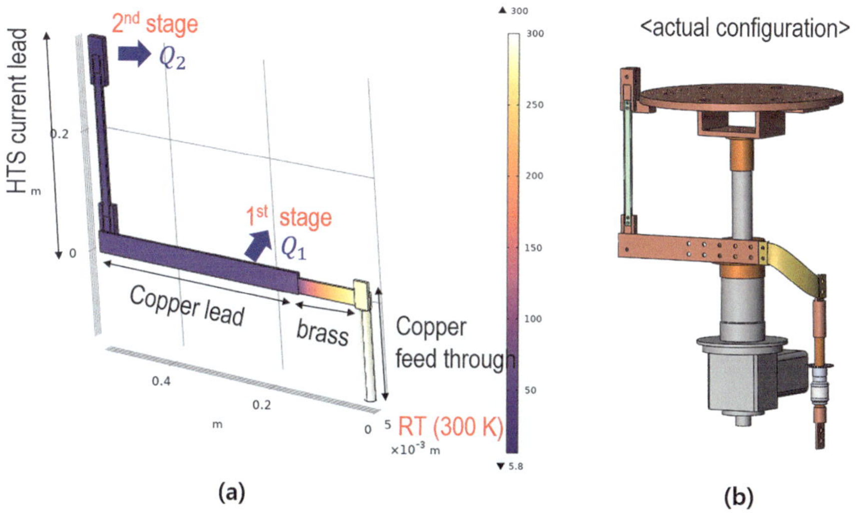

Fig. 3.21 **a** Temperature distributions of current lead calculated by FEM; **b** actual configuration

$$-\vec{n} \cdot \vec{q_c} = \frac{1}{R_c A} \Delta T, \tag{3.17}$$

where q_c, R_c, A, and ΔT are heat flux by the contacts, thermal contact resistance, contact area, and temperature difference between two contacted surfaces. For every thermal contact, thermal contact conductance of 10 W/m^2·K is assumed. Thermal contact conductance highly depends on the applied pressure between the surfaces. In this particular magnet, enough preload is applied to keep its pressure even if the magnet experiences thermal contraction and Lorentz force. Details on the preload structure are discussed in Sect. 4.3.

Figure 3.23 shows temperature analysis results of the WLS magnet when the operating current is 180 A. The estimated second stage temperature is 5.8 K, where

Fig. 3.22 FEM modeling for thermal analysis

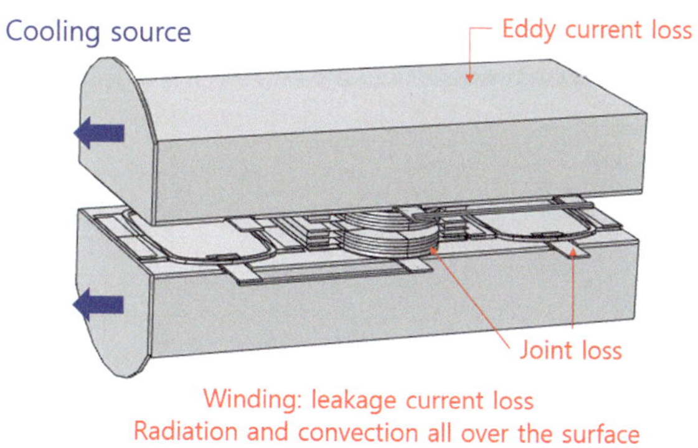

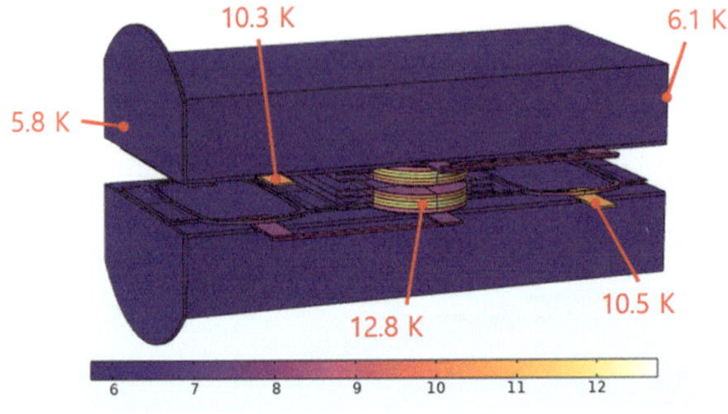

Fig. 3.23 Temperature distributions calculated by thermal FEM model

the maximum temperature is estimated to be 12.8 K at the outer joint. The temperature gradient along the iron yoke is about 0.3 K. Note that the target cooling temperature of the magnet is under 20 K according to the electromagnetic design results.

3.4 Operation Analysis of Iron-Dominated HTS Magnet

3.4.1 Lumped Circuit Model Considering Nonlinear BH Characteristics of Iron Core

The key idea of the NI technique is the elimination of turn-to-turn contact insulation material [59]. The major advantages of the NI technique are increased engineering current density, higher mechanical integrity, and self-protection behavior against electrical burnout. However, the current bypass mechanism results in the generation of the leakage current during charging the magnet. Magnet charging delay and the additional heat load during charging and discharging are the major obstacles to the NI technique [73, 96].

The electrical circuit diagram of the lumped circuit model may be represented by Fig. 3.24 [59]. I_{op}, I_θ, I_r, R_{sc}, L, and R_c are power supply current, current flowing in the winding direction, leakage current, superconductor resistance, inductance, and contact resistance. The inductance of the winding and resistance of the superconductor are series connected in the current flowing direction, while contact resistance is parallel connected [73, 96, 97]. The resistance of the superconductor may be represented by the index (n) law [98].

$$R_{sc} = \frac{E_c l}{I_c}\left(\frac{I_\theta}{I_c}\right)^{n-1}, \tag{3.18}$$

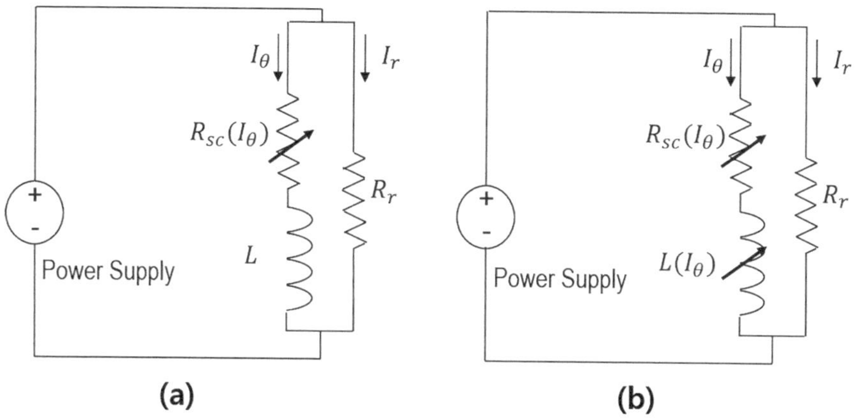

Fig. 3.24 Lumped circuit model of NI coils: **a** air core; **b** iron core

where l and E_c are the length of the conductor and reference electric field criteria. Electric circuit relations may be given by Eq. (3.19).

$$\begin{aligned} V &= L(I_\theta)\frac{dI_\theta}{dt} + R_{sc}I_\theta = R_c I_r \\ I_{op} &= I_\theta + I_r. \end{aligned} \tag{3.19}$$

Characteristic resistance of the NI magnet can be evaluated by

$$R_c = \sum_{j=1}^{N_j}\sum_{i=1}^{N_{ij}} \frac{R_{ct}}{l_{ij}w_j}, \tag{3.20}$$

where R_{ct}, w_j, N_j, N_{ij}, and l_{ij} are, respectively, contact resistance (units: μ Ω·cm^2), the width of conductor at j^{th} PC, number of PCs, number of turns at j^{th} PC, and conductor of ith turn at j^{th} PC [96]. The inductance of iron-core magnet depends on the operating current since the permeability of iron changes as a function of the coil current. Inductance of a nonlinear system can be modeled by the energy method [99].

$$L = 2\frac{\partial W_m}{\partial (I^2)}, \tag{3.21}$$

where W_m is the magnetic energy calculated by the FEM analysis model. Figure 3.25 shows the calculated inductance of the designed WLS magnet.

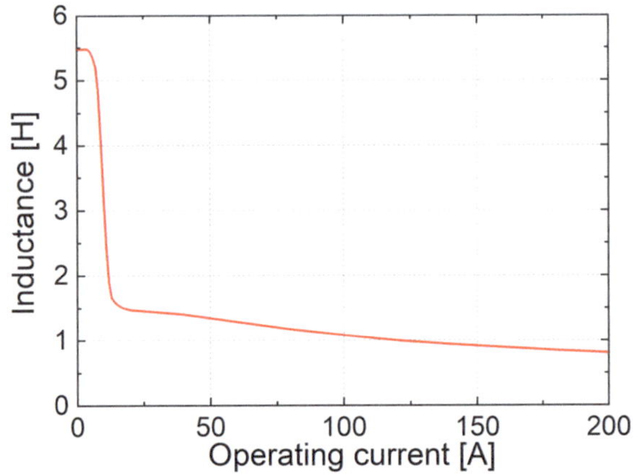

Fig. 3.25 Calculated inductance depending on the operating current

3.4.2 Charging Analysis Results of Metal-Insulated WLS Magnet

Typical ranges of contact resistance of non-insulated coils and metal-insulated coils are 10–100 μ Ω·cm^2 [96, 100–104] and 200–10,000 μ Ω·cm^2 [64, 105, 106], respectively. It is known that the contact resistance varies by applied pressure, surface condition, winding tension, and many other factors. Figure 3.26 compares charging characteristics of the designed WLS magnet assuming R_{ct} of 20 μ Ω·cm^2 and 1,000 μ Ω·cm^2. Linear charging sequence with a maximum ramp rate to keep the leakage current loss under 1 W. Correspondingly, ramp rate is determined to be 8 mA/s and 60 mA/s for R_{ct} of 20 μ Ω·cm^2 and 1,000 μ Ω·cm^2. The estimated charging time in each case is about 40 minutes and 480 minutes, respectively. Considering the working hours of the operator, it is preferable to be fully charged and discharged within a day. Therefore, we decided to design the magnet to be charged within a day by adopting the metal insulation winding technique.

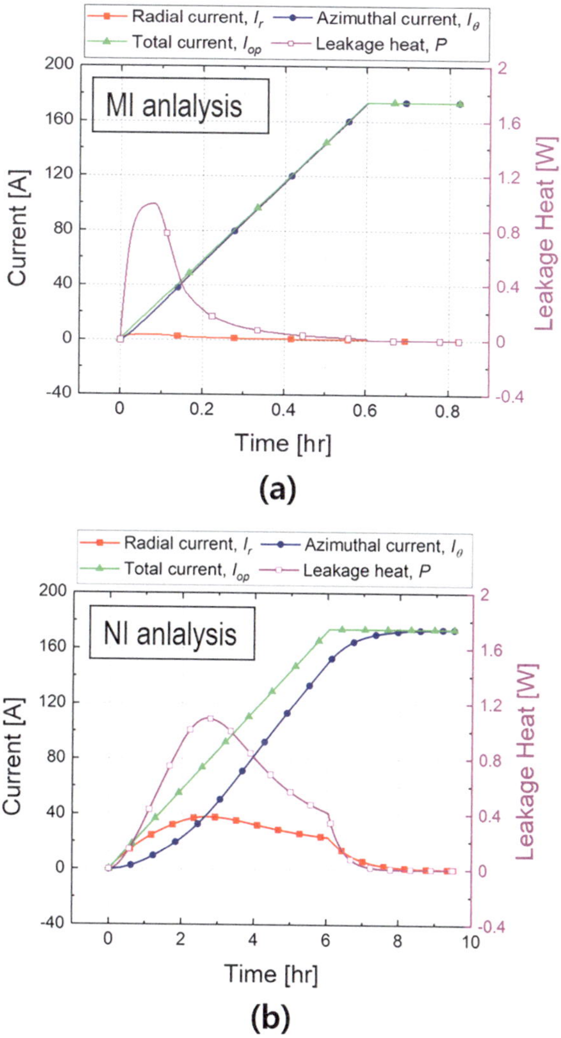

Fig. 3.26 Lumped circuit charging analysis results: **a** R_{ct} of 20 μ Ω·cm^2; **b** R_{ct} of 1,000 μ Ω·cm^2

3.5 Summary of Design Results

Table 3.4 summarizes key parameters of the designed WLS magnet.

Table 3.4 Key parameters of wavelength shifter coil

Main parameters	Values		
REBCO conductor	M4	M5	S4
Manufacturer		SuNAM	
Width [mm]	4.1	5.1	4.1
Overall thickness [μm]	140	120	140
Substrate thickness [μm]		110	
Co-winding thickness [μm]	60	80	60
Magnet dimension			
Magnetic gap [mm]		12	
Cover yoke dimension [cm]		$67\ (W_z) \times 30\ (W_x) \times 10\ (W_y)$	
Height of yoke pole [mm]		30	
Total weight [kg]		360	
Coil dimension			
Conductor width [mm]	4.1	5.1	4.1
Number of pancakes	4×2	1×2	1×4
Number of turns	200	182	40
Length of straight section [mm]	188	188	166
Curvature of straight section [mm]	700	700	700
Curvature of circular section [mm]	25	25	100
Conductor usage [m]	256	116	28
Total conductor usage [m]		1800 (4.1 mm equi.)	
Operation			
Main pole magnetic constant [T/A]	0.11 (20 A); 0.02 (100 A); 0.018 (180 A)		
Side pole magnetic constant [T/A]	0.02 (20 A); 0.018 (100 A); 0.01 (180 A)		
Peak Hoop stress [MPa]		298	
Estimated critical current, avg [A]		210.1 (20 K)	
R_c (R_{ct} = 1,050 μ μ Ω·cm^2) [mΩ]		0.147	
Inductance [H]		5.4-1.5	
Charging time constant [s]		37–10	

Chapter 4
Core Manufacturing Technologies of HTS WLS Magnet

This chapter describes the core manufacturing technologies of the developed WLS magnet. Core manufacturing technologies for superconductor magnets consist of winding, joint, structure, lead, cryogenics, and operation/evaluation. Some of the technologies are intimately related, and the superconducting magnet manifests its function only if all the elements are present. Failure of any component may lead to the failure of the entire system.

Figure 4.1 shows the final CAD drawing of the HTS WLS magnet considering all the core manufacturing technologies. In this chapter, we introduce how each manufacturing technology is organized. Some of the components adopt conventional manufacturing methods, while several innovative manufacturing for certain components. In particular, we explained the dimension change of winding during the winding process. Also, the concept and design methods of the preload structure are newly suggested.

4.1 Winding: Metal-Insulated HTS Pancake Winding

4.1.1 Winding Configuration, Procedure, and Results

4.1.1.1 Description on Winding Configuration

As described in Sect. 3.4, it is decided to adopt metal-insulated winding technology to reduce charging delay. Considering the availability of winding, 60 μm thickness stainless steel tape is wound together with the REBCO conductor. According to the electromagnetic design results, each main pole consists of one 5 mm width pancake coil and four 4 mm width pancake coils. Each side pole has one 4 mm pancake coil. 4 mm width main pole pancake coils are made by the double pancake coils, while others have inner copper lead. A conductor usage for each double pancake coil is

J. Park, *5 T High Temperature Superconductor 3 Pole Wavelength Shifter Magnet for Accelerator-based Light Sources*, Springer Theses,
https://doi.org/10.1007/978-981-96-8679-7_4

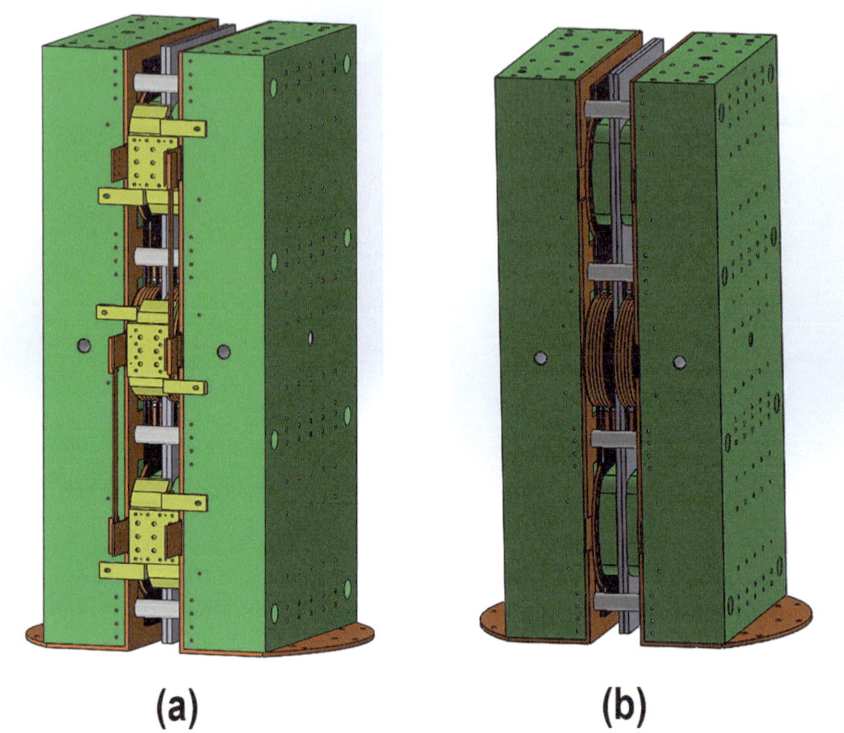

Fig. 4.1 Final CAD drawing of the WLS magnet: **a** front view, **b** back view

256 m, while the available continuous spool length from the manufacturer is under 200 m. Therefore, two splice joints are added for the double pancake coil spools to keep the superconductor face radially outward. For the 5 mm width single pancake coil, the winding starts facing radially inner direction. one splice joint is added. The winding starts facing radially inner direction to solder to the length.

4.1.1.2 Winding Procedure of Double Pancake Winding

The winding procedure of double pancake coil is illustrated in Fig. 4.2.

- Step 1. Division of conductor spool: Considering the conductor usage of 256 m for double pancake coil, half of the total conductor is divided into another winding spool.
- Step 2. Crossover turn installation: The upper and lower pancake coils are electrically connected by a crossover connection at the innermost turn. One winding spool, which is used to wind the upper pancake coil, is fixed on the winding zig to rotate together with the winding mandrel.

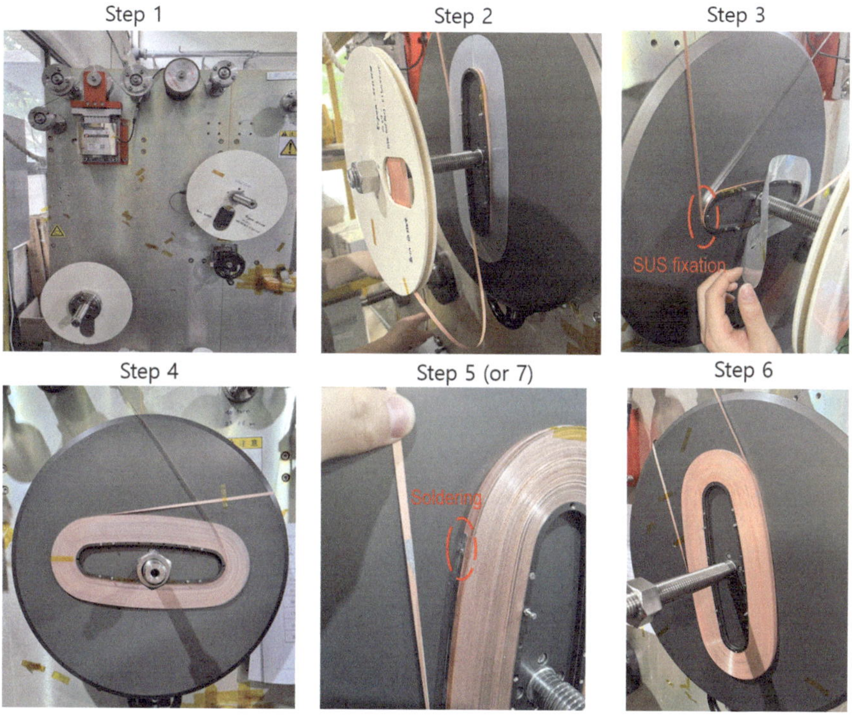

Fig. 4.2 Winding procedure of double pancake coil

- Step 3. Fixation of co-wound stainless tape: Co-wound SUS tape is overlapped by the REBCO conductor. Winding tension applied to the REBCO tape helps to fix the SUS tape.
- Step 4. Co-winding of lower pancake coil: Winding tensions of REBCO and SUS are determined to be 19.6 N and 14.7 N to keep the same strain.
- Step 5. End-turn fixing: The outermost REBCO tape is soldered to the previous REBCO tape.
- Step 6. Co-winding of upper pancake coil: Disassemble the winding spool fixed on the winding zig, and repeat the Step 4 procedure.
- Step 7. End-turn fixing: This procedure is identical to Step 5.

4.1.1.3 Winding Procedure of Single Pancake Winding

The winding procedure of the main pole and side pole single pancake coils is principally the same. The only difference is the total number of winding. The winding procedure of double pancake coil is illustrated in Fig. 4.3.

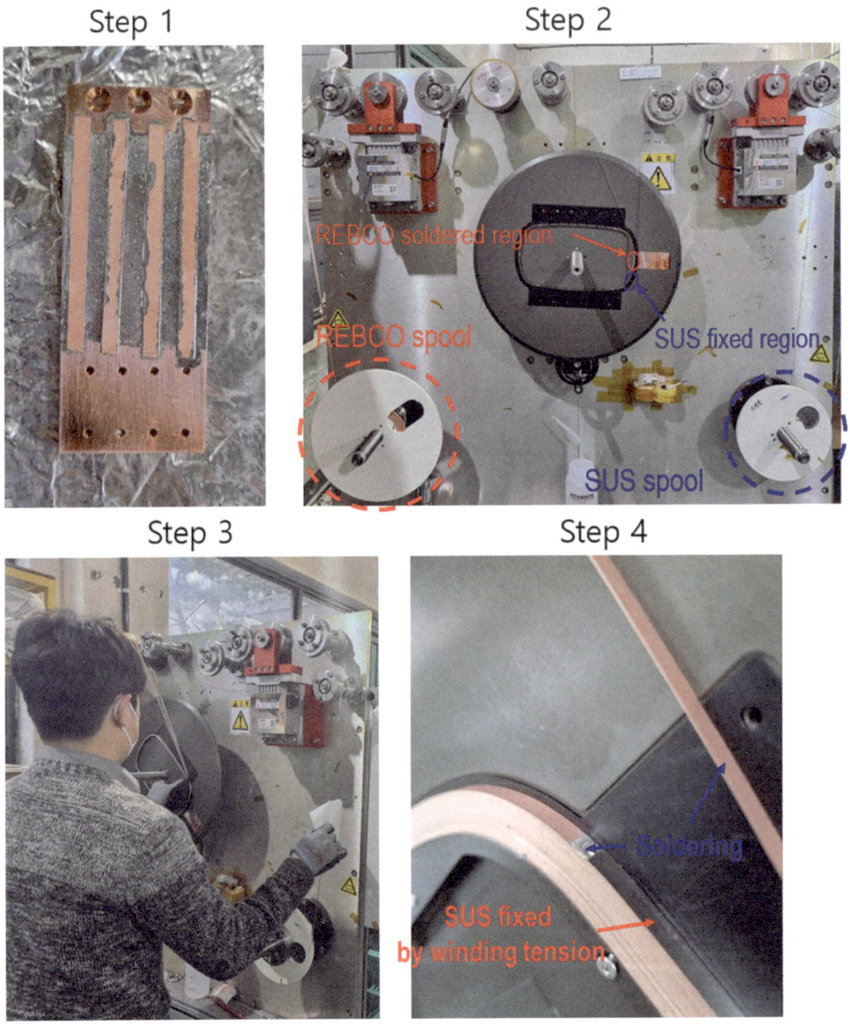

Fig. 4.3 Winding procedure of side pole single pancake coil (identical to the main pole one)

- Step 1. Soldering HTS Conductor on Copper Lead: The superconductor face is soldered to the inner copper lead.
- Step 2. Fixation of Co-wound Stainless Tape: SUS tape is fixed at the aluminum bobbin using the two-sided tape.

- Step 3. Co-winding procedure: The applied winding tension is identical to the double pancake winding.
- Step 4. Fixation of end turn: End-turn fixing method is identical to the double pancake.

4.1.1.4 Observation on Winding Dimension Difference Between Straight and Circular Sections

Figure 4.4 shows a picture of a double pancake coil after winding. The width and height of the double pancake coil are measured. Table 4.1 summarizes the measured dimensions of four wounded double pancake coils. When the relative difference between the width and height of the coil is compared, it is obvious that the thickness of winding is thinner in the straight section winding. In Sect. 4.1.2, a winding stress analysis model to explain this thickness difference is suggested.

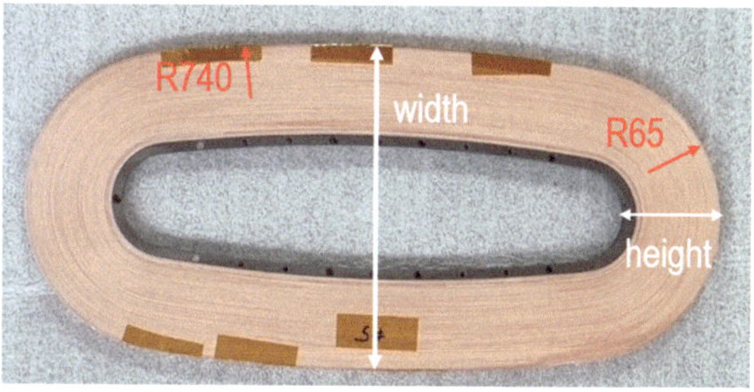

Fig. 4.4 Winding results of main pole double pancake

Table 4.1 Winding dimension measurement results of main pole double pancake coil

		Width [mm]			Height [mm]			Relative difference [mm]
		Designed	Measured	Error	Designed	Measured	Error	
M4#1	Lower	143.2	144.3	1.1	44.5	44.3	–0.2	1.3
	Upper		143.8	0.6		43.9	–0.6	1.2
M4#2	Lower		144.3	1.1		44.1	–0.4	1.5
	Upper		144.2	1.0		44.2	–0.3	1.3
M4#3	Lower		143.4	0.2		43.7	–0.8	1.0
	Upper		143.2	0		43.7	–0.8	0.8
M4#4	Lower		143.3	0.1		44.1	–0.4	0.5
	Upper		143.0	0.2		43.9	–0.6	0.8

4.1.2 *Suggestion of Winding Dimension Change Analysis Model*

4.1.2.1 Key Suggestions

- **Winding dimension change by winding tension**: Winding tension builds stress distributions within the winding pack and they produce the displacements. By the accumulation of displacements in the winding pack, the total winding dimension may be significantly changed. The change of REBCO PC diameter by the winding tension has yet to be reported. Here, we report the aforementioned issue together with the formulas to calculate winding dimension change.
- **Winding compressiblity**: Recently, compressible characteristics of REBCO contacts have been studied for stacked sheets and dry wound pancake coil [107, 108]. Xue et. al measured stress-strain behavior of stacked REBCO sheets, and they measured very low stiffness for the stacked sheets as well as the nonlinear stress-strain behavior [107]. Yan *et al.* explained low radial stress build-up of dry wound pancake coil by both compressibility and asperity contact theory which also indicate very low radial Young's modulus of dry wound winding pack [108–112]. Low stiffness of dry wound PC during winding process may induce large change of radial dimension.

4.1.2.2 Theoretical Modeling

Figure 4.5 shows a schematic of dimension change during the winding process [113]. a_i, Δr_i, and δu_i are expected winding dimension of ith turn without winding tension, radial dimension change when i turns are wounded, and incremental displacement in i^{th} winding procedure. Exceptionally, a_0 is the inner radius of the bobbin. The winding stress distribution of the concentric cylinder is calculated by incrementally applying a winding tension load. Analytic solution for incremental displacement (δu_i), radial stress ($\delta\sigma_{r,i}$), and hoop stress ($\delta\sigma_{\theta,i}$) by i^{th} turn winding can be obtained by solving force balance equation [114]:

$$\begin{aligned} \delta u_i(r) &= C_i r^k + D_i r^{-k}, \\ \delta\sigma_{r,i}(r) &= \frac{C_i}{kL - V} r^{k-1} - \frac{D_i}{kL + V} r^{-k-1}, \\ \delta\sigma_{\theta,i}(r) &= \frac{kC_i}{kL - V} r^{k-1} + \frac{kD_i}{kL + V} r^{-k-1}, \end{aligned} \tag{4.1}$$

where r is the radial position. k, L, and V are represented by

$$k = \left(\frac{R}{L}\right)^{0.5}, L = \frac{1}{E_\theta}, R = \frac{1}{E_r}, V = \frac{\nu}{E_\theta}, \tag{4.2}$$

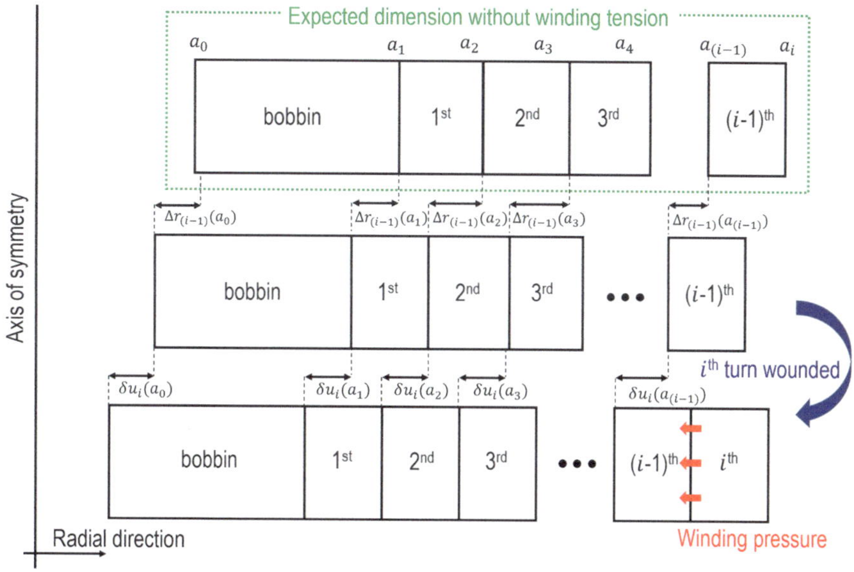

Fig. 4.5 Axisymmetric schematic view of radial dimension change of solenoid pancake coil during winding procedure

where E_θ, E_r, and ν are hoop, radial direction Young's modulus, and Poisson ratio. C_i and D_i can be obtained by solving the boundary condition:

$$\delta\sigma_{r,i}(r = a_0) = 0, \delta\sigma_{r,i}(r = a_i) = \frac{Td}{a_i}, \tag{4.3}$$

where T and d represent winding tension pressure and conductor thickness. Then, the radial dimension change of jth winding turn when i^{th} turn is added may be represented by

$$\begin{aligned} \Delta r_i(a_j) &= \Delta r_i(a_{j-1}) - \frac{\nu T}{E_\theta} d \quad (j = i) \\ \Delta r_i(a_j) &= \Delta r_{i-1}(a_j) + \delta u_i(a_j) \quad (j < i). \end{aligned} \tag{4.4}$$

The thickness change of the initial thickness change of the tape by winding tension may be represented by the Poisson ratio relation. Finally, radial dimension change may be obtained by

$$\Delta r_i(a_i) = \Sigma_{i=1}^{j} \delta u_i(a_j) - \frac{\nu T}{E_\theta} jd. \tag{4.5}$$

4.1.2.3 Experimental Verification

Figure 4.6 shows a picture of a winding zig and bobbin fixed on the winding machine. A total of 150 turns of SS tape, whose thickness is 140 um, are wound on the 8-mm-thick ring-shaped aluminum bobbin. The winding bobbin consists of two 4 mm aluminum rings, and the first turn is fixed at the front ring by the two-sided tape, and the next turn is placed on the back ring by crossover connection [115]. One edge of the conductor is placed against the winding zig and tapes are wounded on the back aluminum ring. At every 30 turn winding, the outer diameter of the winding pack is measured by the Vernier Calipers with a resolution of 0.01 mm. Outer diameter at 4 different circumferential points were measured and averaged out to estimate winding diameter. To eliminate lengthwise thickness variation of SS or REBCO tape, identical long SS or REBCO tape is repeatedly wound by different winding tensions starting from 9.8 N to 88.2 N with an interval of 19.6 N. During the unwinding process, winding tension is maintained to be 9.8 N. This cycle was repeated 3 times for each tape, therefore, each PC experienced a total of 15 winding and unwinding processes. Table 4.2 summarizes key parameters of the dry wound pancake coil. The thickness of SS and REBCO tape might be slightly different from what was ordered from the manufacturer, since the winding tension may flatten the tape geometry which will be discussed in the latter section.

Figure 4.7 illustrates winding tension-dependent radius change of 150 turns SS and REBCO PCs with respect to the radius of PC wound by 9.8 N winding tension.

Fig. 4.6 Aluminum winding zig and bobbin fixed on the winding machine

Table 4.2 Key parameters of the dry wound pancake coil

Parameters	Values
Dimension	
Bobbin inner radius; outer radius [mm]	52; 60
Winding inner radius; outer radius [mm]	60; 81
REBCO and SS tape thickness; width[mm]	0.14; 4.1
SS tape Young's modulus[GPa]	195
REBCO tape Young's moduli, E_r; E_θ; E_z [GPa]	169; 182; 182
Winding tension (1 kgf=17.1 MPa) [kgf]	1, 3, 5, 7, 9
Total conductor usage [m]	68.2

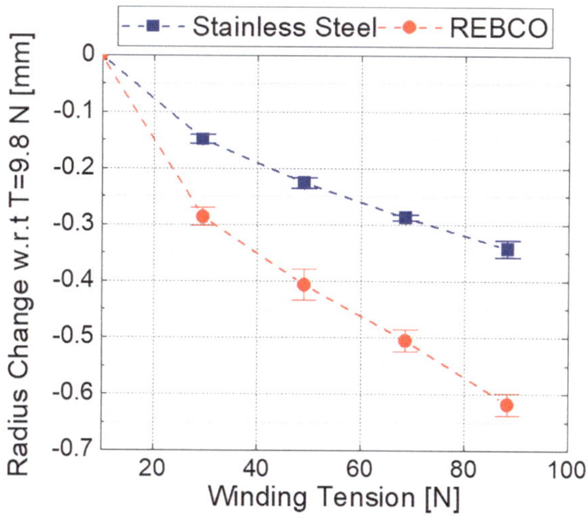

Fig. 4.7 Radius change of 150 turn SS and REBCO pancake coils with respect to the PC wounded by 9.8 N winding tension

Measurement results at the first cycle are only shown. Error bar indicates the standard deviation of measured radius difference on four different circumferential points shown in Fig. 4.6. Radius changes between 9.8 N and 88.2 N winding tension are 0.34 mm and 0.62 mm, respectively, for SS and REBCO coil. A larger radius change is expected for the REBCO PC, this is mainly because the radial Young's modulus is smaller compared to that of SS. According to Eq. (4.1–4.5), radial dimension change linearly increases by the winding tension if Young's modulus is not changed. However, the slope of the measured radius change with respect to winding tension decreases as the winding tension increases. This result implies effective Young's modulus of the winding pack changes depending on the winding tension, which might be explained by the compressible characteristics of rough surfaces in contact.

Figure 4.8 shows the measured radius change of PCs at every 30 turns with respect to the radius of PC wound by 9.8 N winding tension. The dashed and solid line shows the radius change of PC calculated by Eq. (4.5) using the original radial Young's modulus and the modified values, respectively. The modified Young's modulus is deter-

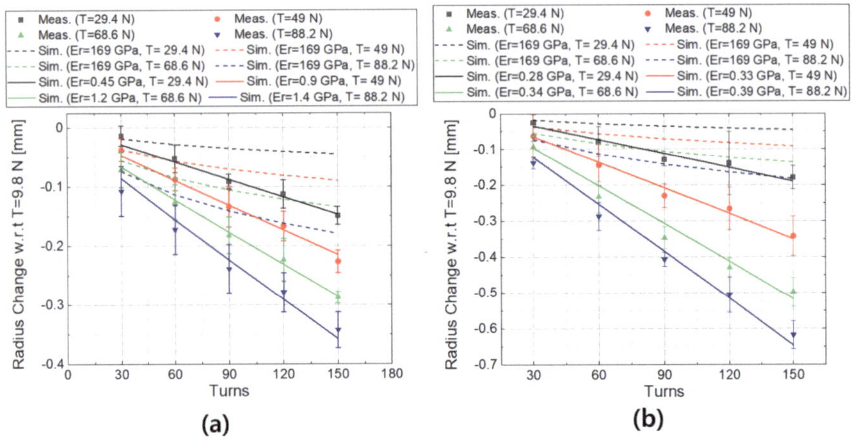

Fig. 4.8 Measured radius change of **a** SS and **b** REBCO PC with respect to the one wound by 9.8 N winding tension

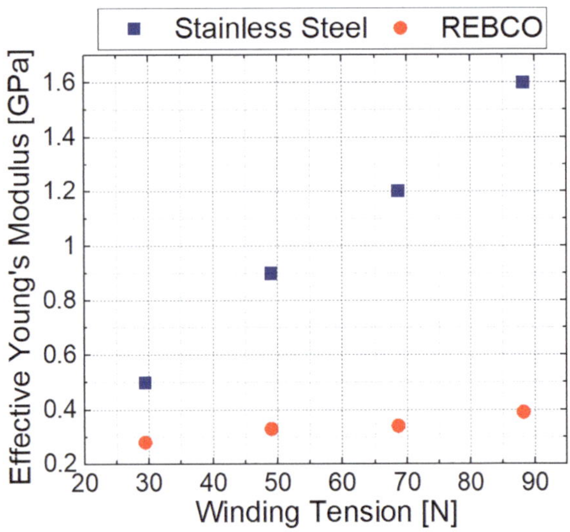

Fig. 4.9 Effective Young's modulus to best explain the measured radius change

mined to minimize the mean square error between the measured and simulated radial change. In this calculation, it is assumed that the radial Young's modulus depends on the winding tension. Figure 4.9 shows the effective radial Young's modulus to best fit the simulation and measurement radius change depending on the winding tension. Effective Young's modulus ranges from 0.5 to 1.6 GPa and 0.28 to 0.39 GPa, respectively, for SS and REBCO PCs. These values are significantly lower than the material's original Young's modulus, where this low value may be explained by the compressible characteristics of rough turn-to-turn contacts as studied in the previous literature [108]. Figure 4.10 shows radial displacement and radial stress calculation

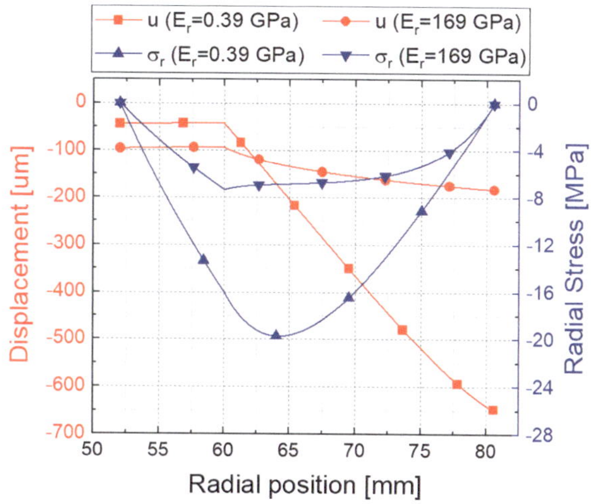

Fig. 4.10 Radial displacement and radial pressure calculated for REBCO PC with 88.2 N winding tension applied

results for 88.2 N winding tension applied for REBCO PC. Maximum radial stress of 19.7 MPa was estimated for the original REBCO's Young's moduli, while 7.3 MPa for the modified one. In the bobbin region, the displacement is almost uniform, while displacement in the winding significantly increases with the radius since the displacement of the winding is accumulated at every winding step.

4.1.2.4 Observation on Irreversible REBCO Winding Change

Figure 4.11 shows the measured average thickness of REBCO tape wound with different winding tension. One cycle includes five winding and unwinding procedures, where winding tension during winding is increased from 9.8 N to 88.2 N with an interval of 19.6 N at each procedure while winding tension during unwinding cycle is fixed to 9.8 N. The average thickness of REBCO is evaluated by the average thickness of the winding pack divided by the total number of turns. The error bar indicates the standard deviation of measured thicknesses. Unlike SS PC winding, the irreversible radial dimension change of REBCO PC is observed during the consecutive winding cycles. The average thickness of the tape is decreased by 1.3 μm and 0.5 μm (2.0 mm and 0.8 mm in radius change), when 9.8 N winding tension is applied. A gradual decrease of thickness change can be found with higher winding tension and more winding cycles. Especially, the change of thickness is about 0.2 μm for 88.2 N winding tension applied PCs between first and third cycles.

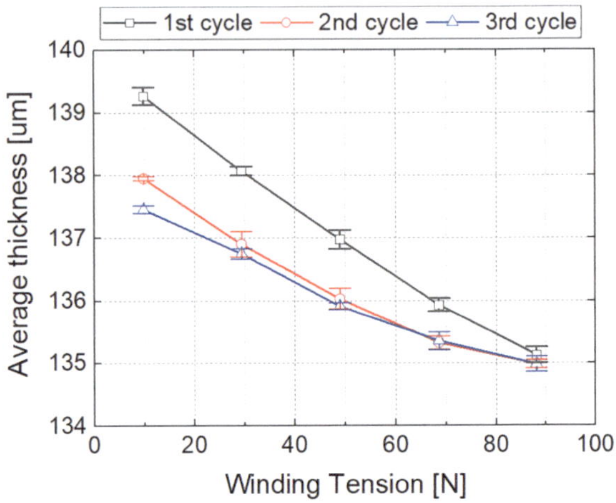

Fig. 4.11 Measured average thickness of 150 turn REBCO pancake coil with consecutive winding cycles

4.1.3 Limitation of Modeling

If the coil geometry is not axisymmetric, the curvature and corresponding winding pressure differ by the circumferential directions. The suggested model enables estimating the relative change of winding dimension due to the explained difference in winding pressure. However, it is rather complicated to estimate the absolute dimension of winding, since the thickness variations of the conductor itself.

As pointed out in the previous studies, REBCO has intrinsic widthwise height difference [116, 117]. The possible explanation for the irreversible thickness variation might be the widthwise flattening of the conductor by winding pressure. In [117], they measured the height difference of REBCO tape after magnet construction and test, and it can be seen that the widthwise height difference is lowered after winding. Initial compression of the REBCO conductor before the coil winding may be helpful to control the radial dimension of dry wound pancake coil, as well as increasing the winding pack density.

4.2 Joint: Resistive Bridge Joint

Figure 4.12 shows the typical topology of the REBCO conductor connection. Additional superconductor tape is soldered to each REBCO tape for bridge-type joints. To reduce the resistivity of the joint, REBCO conductors are soldered while keeping the superconductor sides facing each other. The direction of the superconductor surface is controlled by adding a lap joint in the initial spool as explained in Sect. 4.1.1.

For the pancake-to-pancake connection, the bridge joint is made at the outer radius of the main pole coil. 12-mm-wide REBCO conductors are cut to 8 mm length, and a total of 9 pieces were soldered for each joint. Electrical connection between lower

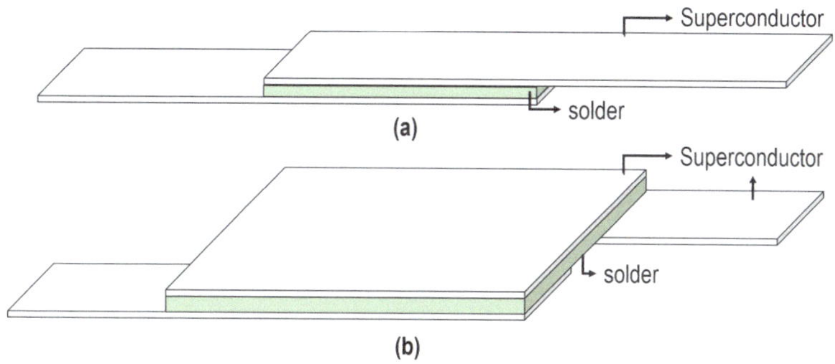

Fig. 4.12 Typical topology of joint: **a** lap joint; **b** bridge joint

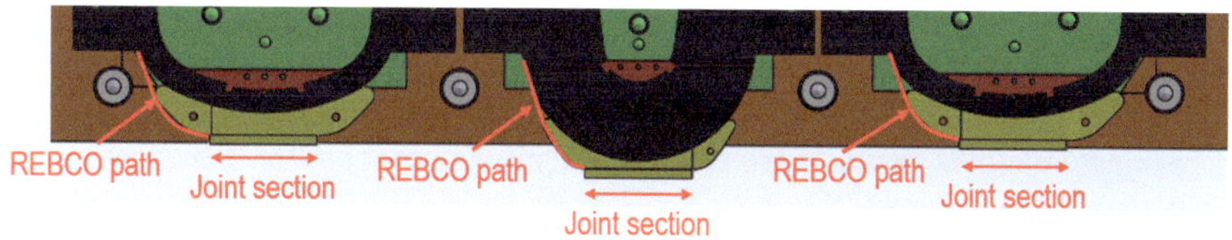

Fig. 4.13 Zig structure for pole-to-pole magnet joint

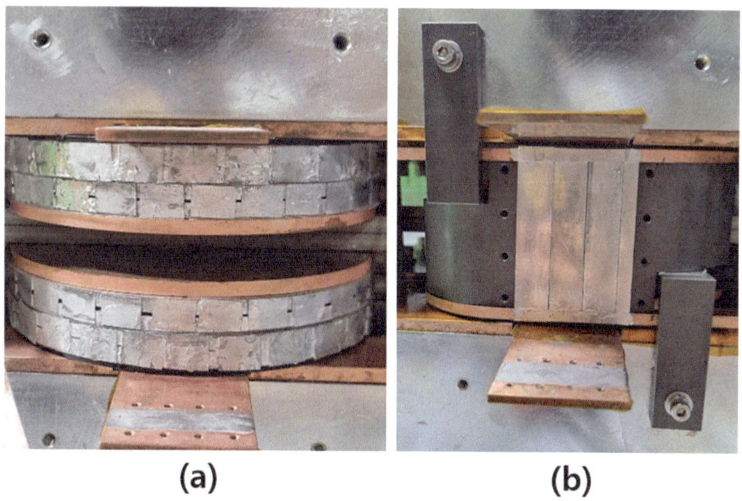

Fig. 4.14 Manufactured resistive joint: **a** pancake-to-pancake joint; **b** side pole joint

and upper pancake coils is performed on the additional structure. Joint structures are individually designed for the side pole and main pole. Figure 4.13 shows the top section of the designed structure. The minimum curvature of the structure is kept under 15 mm to keep the bending strain under 0.4%. The available joint length is 41 mm and 62 mm for side pole and main pole, respectively (Fig. 4.14).

4.3 Structure: Axial Preload Structure

The main role of the preload structure is to maintain the mechanical integrity of the magnet against the deformation by Lorentz force and thermal contraction. Preload is applied in the manufacturing process at room temperature, and the applied pressure should be kept in the cryogenic condition. Another important role of preload structure for dry wound magnets is to keep the contact pressure of every thermal contact. Thermal contact resistance is directly related to the contact pressure, where increased contact pressure is preferable to reduce thermal gradients within the magnet.

4.3.1 Preload Structure for Solenoid Magnets

The concept of preload structure for solenoid magnets is shown in Fig. 4.15. When the magnet is cooled down to the cryogenic temperature, both the magnet and the preload top plate contract. Since the Lorentz force acts in the -z-axis direction, the magnet contracts toward the center of the magnet in the powering process. To keep the magnet compressed with the preload structure, the applied contraction (d) by the initial preload should be larger than the sum of thermal and powering contraction.

$$\Delta d > \Delta z_c' - (\Delta z_c + \Delta z_m), \tag{4.6}$$

where Δd, $\Delta z_c'$, Δz_c, Δz_m are contraction by the preload, deformation of top plate by thermal contraction, that of the magnet by thermal contraction and magnetic force.

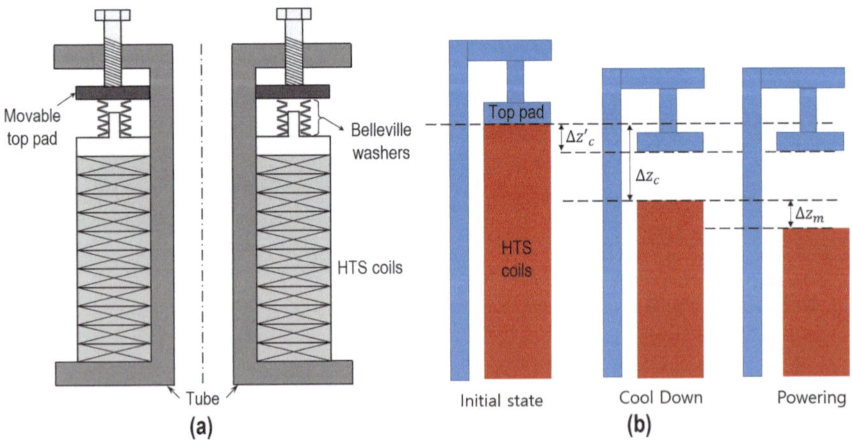

Fig. 4.15 Solenoid preload structure: **a** Schematic of sectional view; **b** deformation by cool down and Lorentz force

4.3.2 *Suggestion of Preload Design Procedure for WLS Magnet*

- **Step 1. Calculate deformation by Lorentz force and cool down**: Solenoid preload structure cannot be applied to the WLS magnet. This is mainly because the system requires center space to install the beam pipe. The upper and lower magnet parts should be fabricated separately, and they should be assembled by some fastening structure. Figure 4.16a shows the schematic of the suggested preload structure for the WLS magnet. Belleville washer is installed between the bolt head and the iron yoke to keep bolt pressure after experiencing the thermal contraction and Lorentz force. Rods are placed in between the upper and lower iron yokes to guide the fastening bolts. Figure 4.16b shows the deformation by cool down and Lorentz force for the suggested preload structure. The following relationship should be satisfied to keep the preload against the aforementioned preload.

$$\Delta d > (\Delta z_c + \Delta z_m) - \Delta z'_c. \tag{4.7}$$

Deformation by the thermal contraction can be simply calculated.

$$\Delta l = \alpha l_0, \tag{4.8}$$

where Δl, α, and l_0 are the change of dimension by thermal contraction, thermal contraction coefficient from 300 K to 4 K, and initial length. Thermal contractions from 300 K to 4 K are 0.24%, 0.30%, 0.32%, and 0.20%, respectively, for GFRP, SUS, copper, and iron. The deformation by the Lorentz force is calculated by the FEM. Figure 4.17 shows the 1,000 times magnified deformation of WLS magnet by the Lorentz force. The half of the main pole coil and iron yoke experience

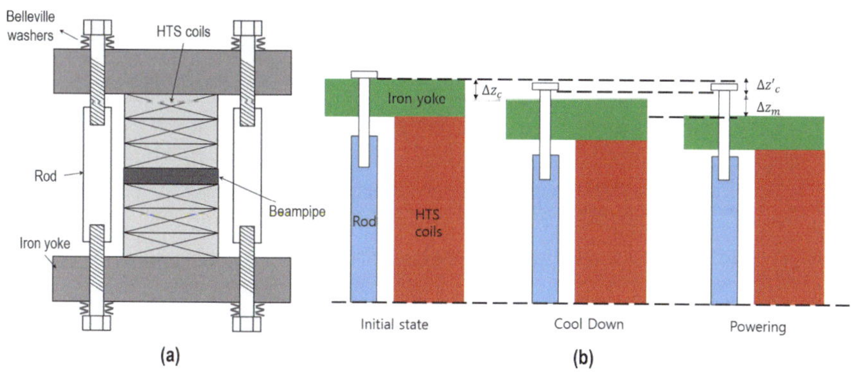

Fig. 4.16 WLS preload structure: **a** Schematic of sectional view; **b** deformation by cool down and Lorentz force

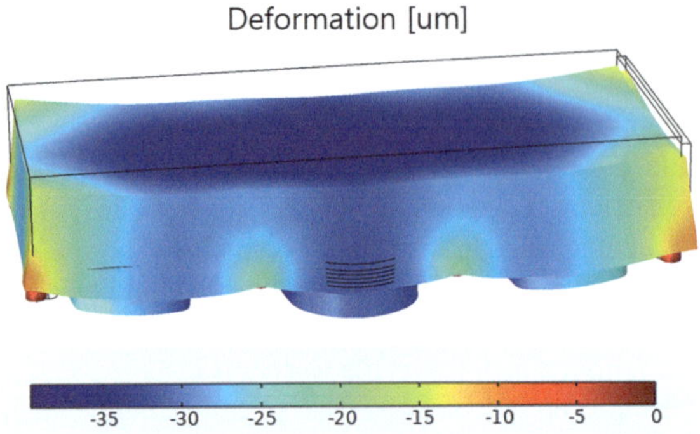

Fig. 4.17 Magnified view of deformation of WLS magnet by Lorentz force

Table 4.3 Calculated deformation of the WLS magnet

Type of load	Iron top (z)	Bolt top(z')	Iron-bolt distance($z - z'$)
Initial status [mm]	0	0	0
Cool down [mm]	–0.31	–0.28	0.03
Energized [mm]	–0.29	0	0.29
Total [mm]	–0.60	–0.28	0.32

46 kN and 252 kN, respectively. The estimated maximum deformation is –38 μm. Table 4.3 summarizes calculated deformation by the cool down and Lorentz force.

- **Step 2. Determine available preload force**: The relation between the bolt fastening torque(T_P) and the fastening preload (F_P) can be formulated by the following:

$$T_P = F_P\left(\frac{p}{2\pi} + \frac{\mu_t r_t}{cos\beta} + \mu_n r_n\right), \tag{4.9}$$

where F_P, p, μ_t, μ_n, r_t, r_n, and β are preload force, pitch length, the friction coefficient of thread, that of the bolt head, the effective radius of the thread, that of the bolt head, and thread angle as shown in Fig. 4.18. The available specification for the preload bolt is M10, where the parameters are summarized in Table 4.4. The maximum fastening torque of the M10 bolt is 65 N·m and the applied fastening torque is determined to be 16 N·m. Finally, the applied preload force is calculated by 6.8 kN per bolt.

- **Step3. Design of Belleville washer configuration**: The Belleville washer is a conical-shaped spring washer with a large spring constant. The total spring constant (k_{tot}) of the Belleville washer configuration should satisfy the given equation.

$$k_{tot} < F_P/\Delta d. \tag{4.10}$$

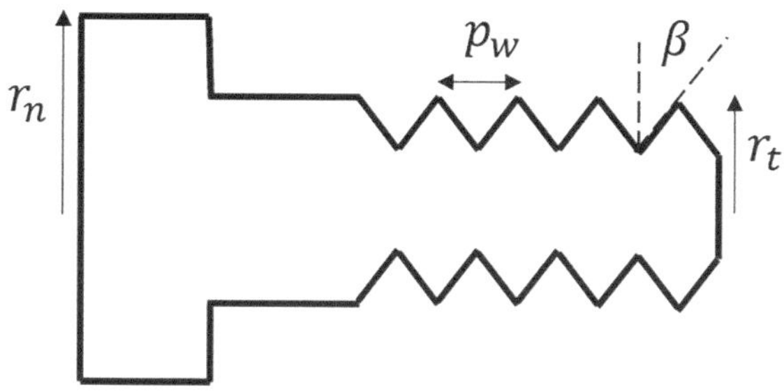

Fig. 4.18 Parameters of bolt thread

Table 4.4 Specification of preload bolt for WLS magnet

Bolt type	p [mm]	β [o]	r_t [mm]	r_n [mm]	μ_t, μ_n
M10	10	30	10	14	0.165

With a combination of series and parallel connection of the Belleville washer, we can create the desired spring constant. The total spring coefficient is

$$k_{tot} = \frac{k}{\sum_{g}^{i=1} \frac{1}{n_i}}, \tag{4.11}$$

where g, n, and k are number of groups, number of washers in ith group, and spring constant of one washer. With three stacks of M10 Belleville washer, the total spring constant is 13 kN/mm. Finally, applied deformation by the preload is calculated by 0.52 mm which is larger than the estimated value of 0.31 mm.

4.4 Cryogenics: Cryogenic System Configuration

4.4.1 Liquid Nitrogen Precooling

It takes a long time to cool down the 400 kg mass of the WLS magnet. Estimated cooling time without any precooling is about 2 weeks for this particular magnet. Here, we installed a liquid nitrogen precooling line to reduce the initial cooling time. The cooling procedure consists of three steps: (1) cool down the magnet around 100 K with liquid nitrogen; (2) pump out LN_2 by using the vacuum pump; and (3) turn on the cryocooler to cool down the magnet to cryogenic temperature. Adiabatic cooldown mode assumes that the magnet exchanges infinitesimal perfect energy exchanges with the cryogen. Under an adiabatic cooldown mode, mass of liquid nitrogen (M_{LN_2}) to cool a magnet is given by

$$M_{LN_2} = M_m \int_{T_w}^{T_c} \frac{C(T)dT}{h_{LN_2}(T) - h_{LN_2}(T_c)}, \tag{4.12}$$

where T_c, T_w, M_m, h, and C represent cryogenic temperature, warm temperature, magnet mass, enthalpy, and heat capacity. Practically, a perfect cooldown cannot be achieved. An extreme case is the dunk mode which assumes the whole magnet is dumped into the cryogen. The required liquid nitrogen mass to cool down the magnet is estimated by

$$M_{LN_2} = M_m \frac{h_{mag}(T_w) - h_{mag}(T_c)}{h_{LN_2}}. \tag{4.13}$$

In an actual situation, a mass of cryogen used for the cooling is in between the dunk mode and the perfect mode. Estimated usage of liquid nitrogen is 97 L and 155 L, respectively, for adiabatic and dunk modes.

4.4.2 Thermal Path for Conduction Cooling Operation

Figure 4.19 shows a schematic of the conduction cooling path of the magnet. The heat generated from the magnet can be extracted to the second stage conduction cooler with the shown path. The detailed view of the cooling channel configuration of the stacked coil is shown in Fig. 4.20. In between the pancake coils, a 0.5-mm-thick aluminum wing is placed to enhance thermal contact between the magnet and to electrically insulate between the stacked coil. OFHC cooling channel contacts the aluminum wing at the side part of each racetrack coil. A cooling plate is attached to

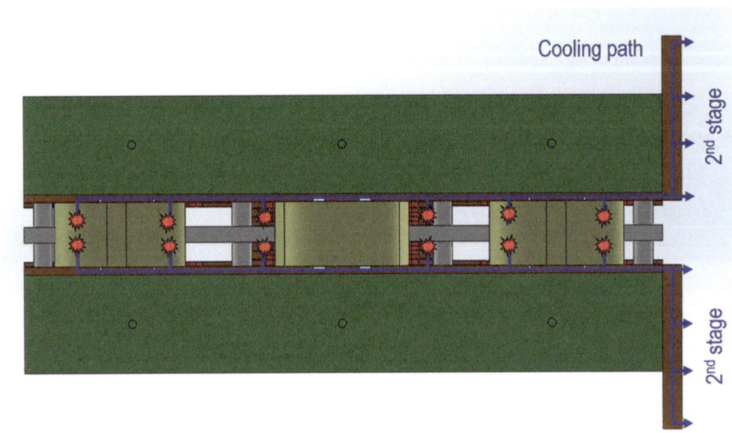

Fig. 4.19 Schematic of conduction cooling path

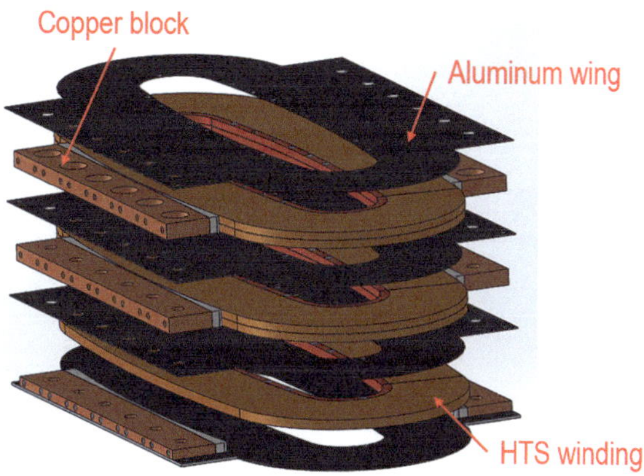

Fig. 4.20 Detailed view of cooling channel for stacked main pole coils

the cover iron yoke, and each pancake coil is stacked on the cooling plate. Finally, one part of the cooling channel is attached to the second stage cooling plate by screwed bolts.

4.5 Lead: HTS Soldered Copper Busbar

For the single pancake coils, REBCO conductor is initially soldered to the inner copper lead. If an additional soldering process is required for the copper lead, the temperature should be controlled so as not to melt the initially attached solder. The use of low melting point solder is a possible solution. In this process, we used two types of solder, wood's metal Sn63Pb37 with melting temperatures of 71 K and 183 K.

The magnet requires two leads to electrically connect the side pole and main pole coils. Figure 4.21 shows the scheme of the electrical connection for every coil. The number indicates the order of the current flow. Between the main pole coil and the side pole coil, the HTS busbar is soldered on the inner copper lead. For the soldering of the HTS busbar on the inner copper lead, wood's metal is used. For the soldering process, the temperature is monitored not to exceed the melting temperature of Sn63Pb37 solder.

4.6 Performance Evaluation: (1) Field (2) Voltage; (3) Temperature

Measuring the physical parameters of the cryogenic system requires dedicated cryogenic instrumentation. In this particular system, we measured the magnetic field, voltage, and temperature of the magnet. The measured magnetic field tells the per-

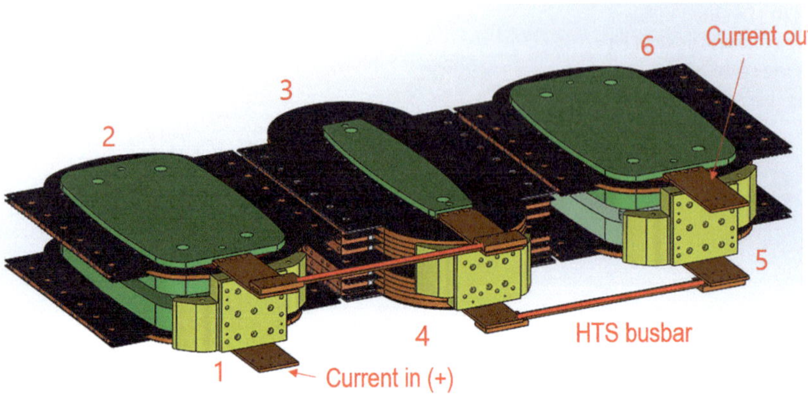

Fig. 4.21 Order of current flow in the WLS magnet

Table 4.5 Specification of instruments for the conduction magnet test

Name	Model	Range
Temperature sensor	Lake Shore DT-670	1.4–500 K
Magnetic field sensor	Lake Shore HGCT3020	> 3 T
Data acquisition	NI SCXI system	–5–5 V
Temperature monitor	Lake Shore Model 218	0–7.5 V
Power supply 1	Oxford mercuryiPS	–10–10 V / -240 A–240 A
Power supply 2	LamdaTDK10-1000	0–10 V/ 0-1000 A
Power supply 3	Lake Shore Model121	100 nA–100 mA

formance of the magnet, while voltage and temperature tell the circumference of the magnet. For NI magnets, voltage gives information on the current distribution within the magnet. Monitoring of temperature is also important to confirm the status of the cryogenic system.

4.6.1 Accuracy of Data Acquisition System and Sensor

The commercial data acquisition system of SCXI 1125 is used for measuring the magnetic field and voltage. Voltage measurement is simply done by the soldered voltage tabs on the magnet. A cryogenic hall sensor with a model name HGCT3020 is used for magnetic field measurement. A silicon diode sensor, DT-670-CU manufactured by Lakeshore, is used to measure the temperature of the magnet. Table 4.5 summarizes the accuracy of the sensors and data acquisition system.

4.6.2 *Instrumentations for Measurement*

4.6.2.1 Voltage

Figure 4.22 shows the location of installed voltage tabs. Every voltage of coil and joint is measured, where a total of 20 voltage tabs are installed. The voltage limit settings for the joint, individual coils, and magnet terminal are 10 mV, 50 mV, and 200 mV, respectively.

4.6.2.2 Magnetic Field

Hall sensor zig is installed at the center of the magnet, which is originally the place for beam pipe installation. Four Hall sensors, two for main pole and one for each side pole, are installed, but Hall sensor signals at the side pole are lost. The only available field data is at the main pole. Two Hall sensors installed at the main pole have a distance of 20 mm. Cross-sectional drawing of Hall sensor zig is shown in Fig. 4.23.

4.6.2.3 Temperature

A total of 11 temperature sensors are installed as shown in Fig. 4.24. Four sensors on the left part of Fig. 4.24 are to obtain information on the cryostat while the other

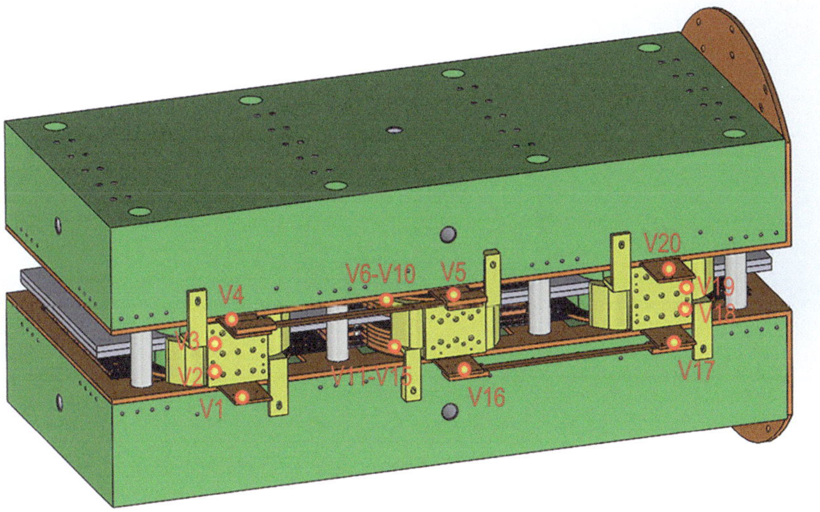

Fig. 4.22 Location of installed voltage tabs

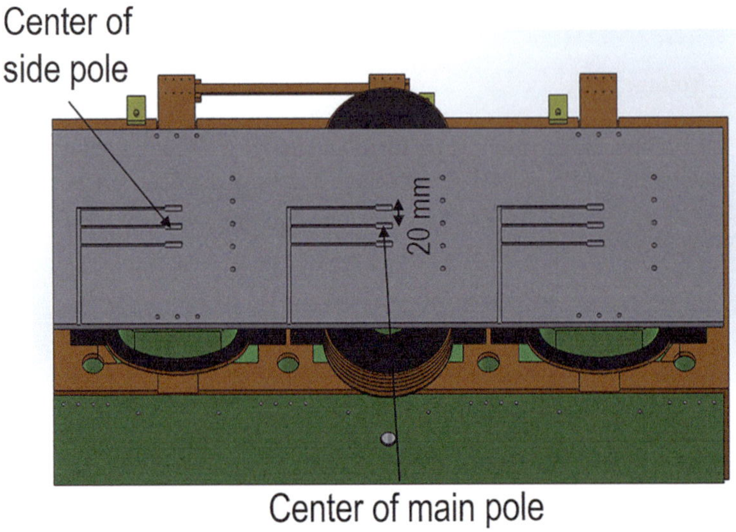

Fig. 4.23 Location of installed hall sensors

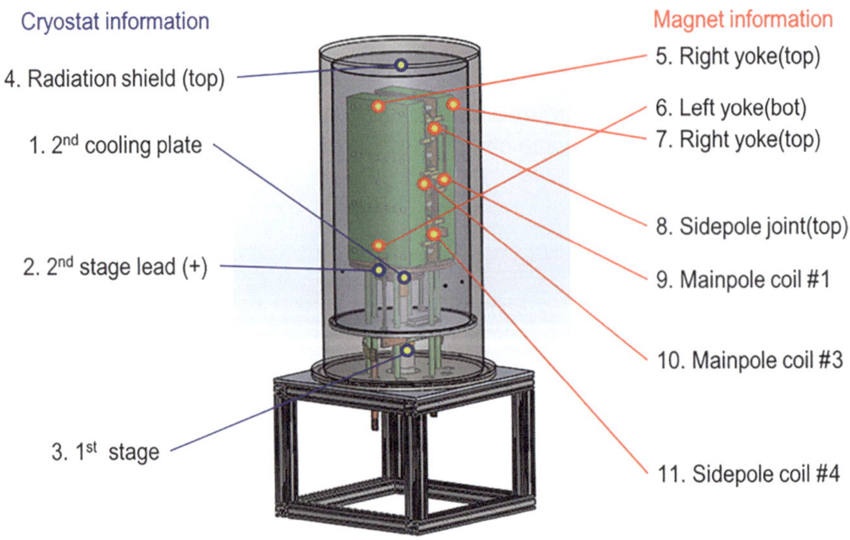

Fig. 4.24 Location of installed temperature sensors

seven are monitoring the temperature of the magnet. Three of them are installed at the iron yoke, while the rest of them are fixed at the aluminum wing of the coils.

Chapter 5
Construction of HTS WLS Magnet

This section describes the construction process of the WLS magnet which incorporates considerations in the design process and developing core manufacturing technologies. Fabricated metal-insulated HTS coils are individually tested in liquid nitrogen to confirm their performance. After that, every pancake coil is stacked and assembled on the cover iron yoke. Finally, the magnet is installed at the in-house conduction cooling test facility.

5.1 Fabrication of Metal-Insulation HTS Racetrack Coils

Figure 5.1 shows all of the fabricated coils, in total 10 pancake coils. Each PC is individually tested in the liquid nitrogen circumference after the winding to confirm the manufacturability. Figure 5.2 shows an example of LN2 test configuration. A Hall sensor is placed at the center of the coil, and the crimp terminals electrically connect the coil. Voltage taps are installed to measure the terminal voltage of the coil. Two types of tests are performed for the magnet, a charging-discharging test to measure the critical current and a fast discharging test to measure contact resistance. Figure 5.3 shows the measured voltage, field, and current of S4#3. For the charging-discharging test, the coil is ramped up to 100 A which is larger than the critical current of the magnet. Repeated charging sequences with a ramp rate of 0.2 A/s, 0.5 A/s, and 1 A/s are performed. Otherwise, the coil is ramped down at 5 A with the maximum allowable ramp rate for the fast discharging test.

Characteristic parameters of individual coils are obtained by comparing the simulation and the measurement data. The coil critical current is determined with the 0.1 μV/cm criteria. With a lumped circuit model described in Sect. 3.4, the decaying of the magnetic field depending on the contact resistance can be simulated when

J. Park, *5 T High Temperature Superconductor 3 Pole Wavelength Shifter Magnet for Accelerator-based Light Sources*, Springer Theses,
https://doi.org/10.1007/978-981-96-8679-7_5

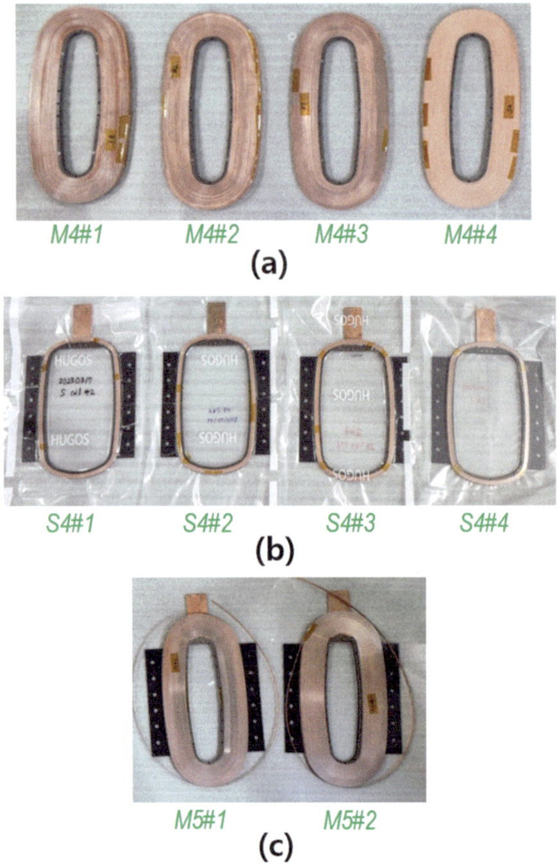

Fig. 5.1 Fabricated metal-insulated coils: **a** main pole DP coils; **b** side pole SP coils; **c** main pole SP coils

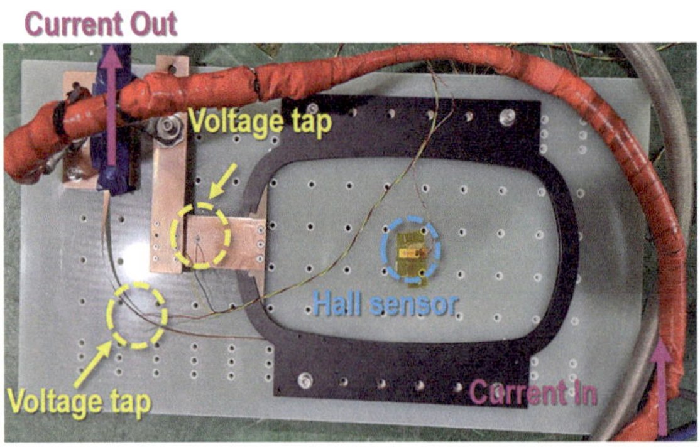

Fig. 5.2 LN2 test configuration for side pole coils

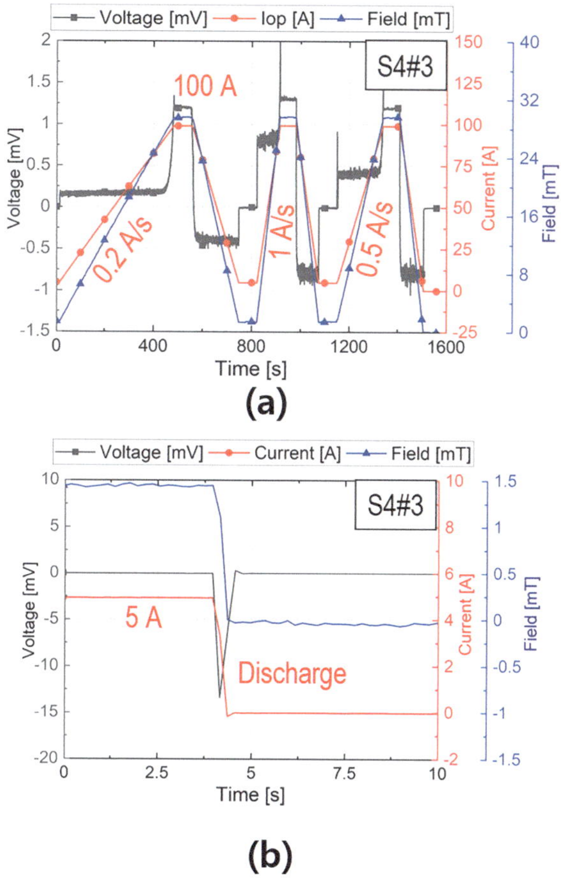

Fig. 5.3 Test results of S4#3: **a** charging-discharging test; **b** fast discharging test

Table 5.1 Performance inspection results of coils at LN2

	Side pole				Main pole		Main pole			
	S4#1	S4#2	S4#3	S4#4	M5#1	M5#2	M4#1	M4#2	M4#3	M4#4
R_{ct} (mΩ·cm^2)	33.6	24.0	9.5	18.8	40.8	41.3	34.4	24.5	23.9	29.4
R (nΩ)	770	555	318	1396	1424	1009	NA	NA	NA	NA
I_c (A)	103.0	102.3	95.8	102.0	103.1	110.0	63.6	63.3	64.0	61.1

the power supply current profile is given. The contact resistance of the coil is calculated to minimize the error between the simulated and measured magnetic field. Table 5.1 summarizes the performance inspection results of the coil. Test results of individual pancake coils are summarized in Appendix.

5.2 Magnet Assembly and Test System Construction

5.2.1 *Order of Coil Stacking*

Based on the critical current measurement results of pancake coils, the order of coil stack is determined. Figure 5.6 shows the stacking order of manufactured coils to maximize the critical current of the magnet. Since the critical currents of M4#2 and M4#3 are larger than the others, these coils are placed near the magnet center.

5.2.2 *Assembly Process*

Figure 5.4 shows the assembly process of the magnet. Winding and individual performance test procedure are explained in the previous section. The cooling plate is assembled on the iron yoke, and HTS coils are stacked on the cooling plate. The stacking sequence is explained in Fig. 4.20. After that, the upper and lower parts of the magnet are assembled using the preload rod. Designed Belleville washer configuration and fastening torque is applied as explained in Sect. 4.3. After that, joint and lead are fabricated for the electrical connection of the coil as explained in Sects. 4.2 and 4.5.

Figure 5.5 shows the assembled magnet on the in-house conducting cooling system without a radiation shield and vacuum chamber. The installed precooling pipe allows the liquid nitrogen to be released at the top of the magnet. Voltage measurement rails

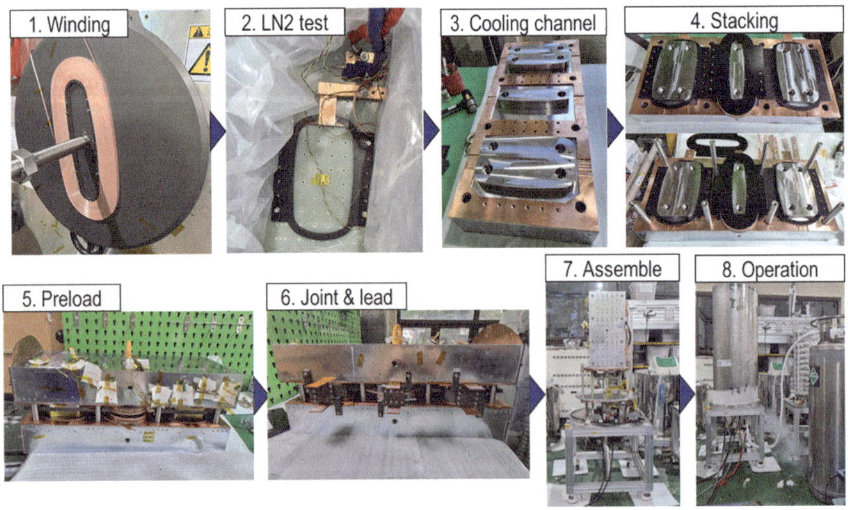

Fig. 5.4 Assembly procedure of the WLS magnet

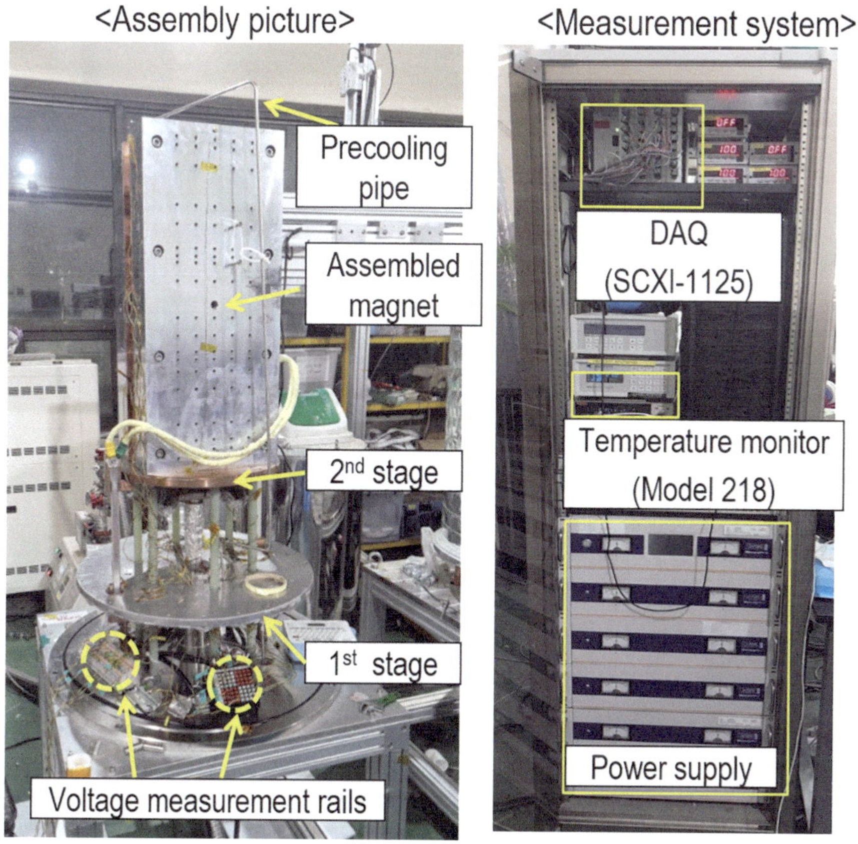

Fig. 5.5 Assembled picture of magnet and measurement system configuration

are placed at the bottom of the cryostat and the signals are connected to the room temperature monitoring system through the vacuum feedthrough. The magnet is placed on the top of the second stage cooling plate.

5.2.2.1 Field Distortion by Assembly Error

Coil stacks and pole yokes are assembled on the cover yoke. Alignment pins are used to assemble the coil stacks and pole yokes, therefore, the structures are aligned with respect to each cover yoke. However, there might be misalignment between the top and bottom cover yokes, since the use of an alignment pin is not available in the current design. The difficulty on top and bottom cover yoke alignment came from the installation of the beampipe which should be positioned in between the top and bottom cover yokes. For example, the designed magnetic gap was 12 mm, but the measured magnetic gap after the preload assembly was 12.4 mm. Figure 5.7 shows the

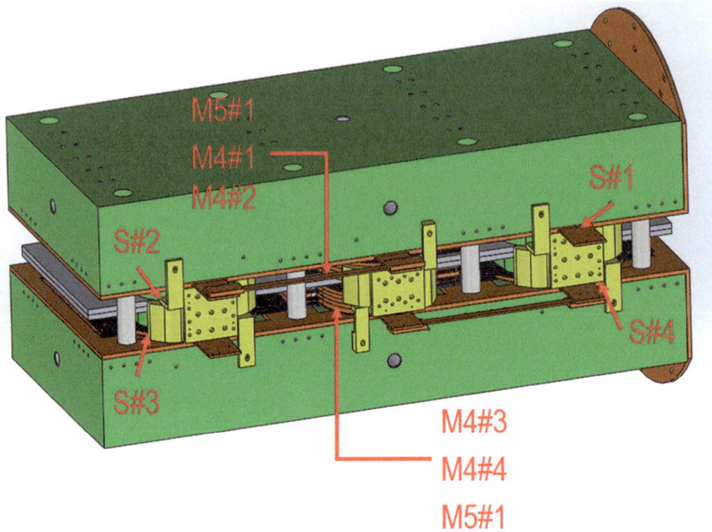

Fig. 5.6 Coil stacking order of manufactured coils

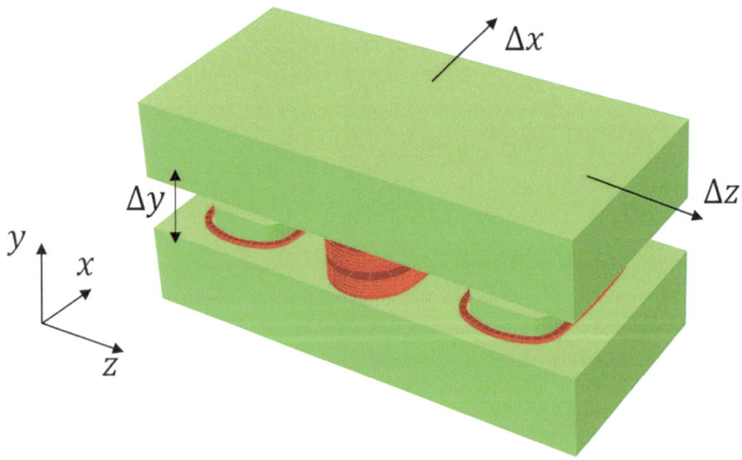

Fig. 5.7 Misalignment of magnet assembly investigated in this study

possible misalignment that may occur during the assembly process. Field uniformity distortion may happen due to the misalignment of the magnet structure. To investigate the misalignment effects on the magnetic field distributions, we assumed three different misalignment situations that the top and bottom cover yokes are misaligned 100 μm in *x*-, *y*-, and *z*-directions. Additional dimension errors may also come from the thermal contraction, Lorentz force-induced deformation, and dimension error of fabricated parts.

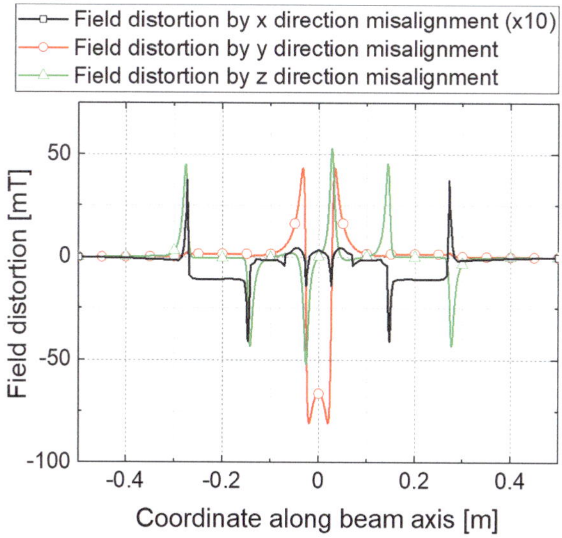

Fig. 5.8 Magnetic field error by misalignment

Table 5.2 Simulated field change by misalignment

Direction	Magnitude [μm]	Field integral (10^{-6} T·m)	Main pole field (mT)
x	+100	−60	0.4
y	+100	−12	−6.6
z	+100	36	0.1

Figure 5.8 shows magnetic field simulation results when the aforementioned alignment error occurred. All three alignment errors change the first field integral uniformity which is related to the deflection angle of the electron beam. The calculated changes of the main field, first field integral, and main pole field in each situation are summarized in Table 5.2.

Chapter 6
Conduction Cooling Test Results of HTS WLS Magnet

This section describes the conduction cooling test results of the HTS WLS magnet. We measured the voltage, temperature, and field of the magnet during the powering process. Comparing the simulation and the measurement, we validate our modeling methods. We observed a reduction in the main pole field mainly due to the screening current-induced field. We suggest a screening current analysis model considering the nonlinear B-H characteristics of iron core based on the 2D H-formulation. Based on the current simulation results, we introduce the segmentation method to calculate 3D field distributions to calculate change of beam property by the screening current-induced field.

6.1 Initial Cooling History

Precooling with liquid nitrogen has been implemented to reduce the initial cooling time of the magnet. Figure 6.1 shows the measured temperature distributions in the initial cooling history. It took 10 h to cool down the magnet to reach 100 K with precooling, and 32 h to reach the final temperature from 100 K. A total of around 110 L of liquid helium is used for the precooling, where the estimated value is between 97 L and 155 L as described in Sect. 4.4.1. The final reached temperature is summarized in Table 6.1.

J. Park, *5 T High Temperature Superconductor 3 Pole Wavelength Shifter Magnet for Accelerator-based Light Sources*, Springer Theses,
https://doi.org/10.1007/978-981-96-8679-7_6

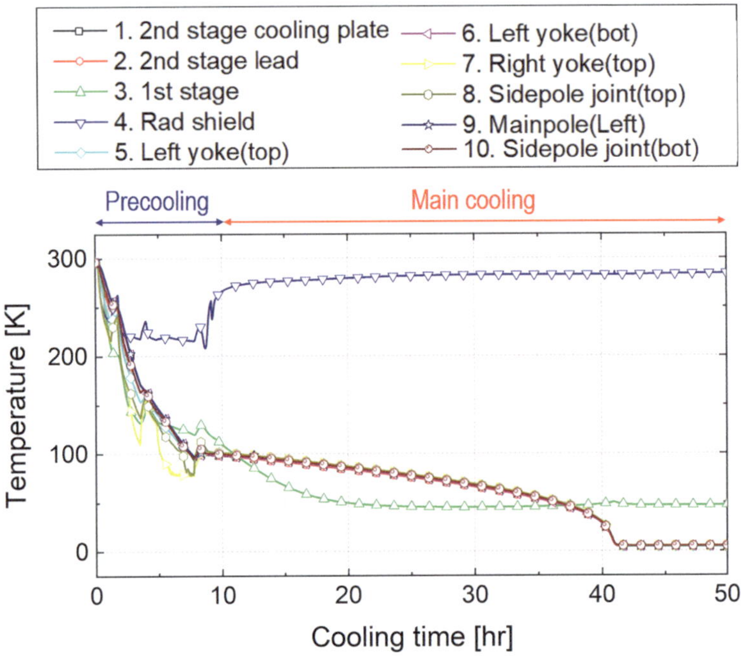

Fig. 6.1 Initial cooling history of the WLS magnet

Table 6.1 Final reached temperatures at initial cool down

Location	1	2	3	4	5	6	7	8	9	10	11
Temperature [K]	4.5	5.4	47.6	284.4	4.6	4.8	4.9	5.5	5.1	4.6	5.3

6.2 Main Powering Test: Generation of 5 T Center Field

The magnet is ramped up to an operating current of 130 A with a ramp rate of 1 A/min and successfully reached a 5 T center field. During the ramp-up, voltage, temperature, and magnetic fields are measured. Here, we explain the details of the measured data together with the simulated values.

6.2.1 Temperature Measurement Results

Figure 6.2 shows measured temperatures at the second stage of the magnet during charging and discharging the magnet. With an operating current of 130 A, the temperature of second stage cooling plated is increased from 4.37 K to 5.2 K. Additional

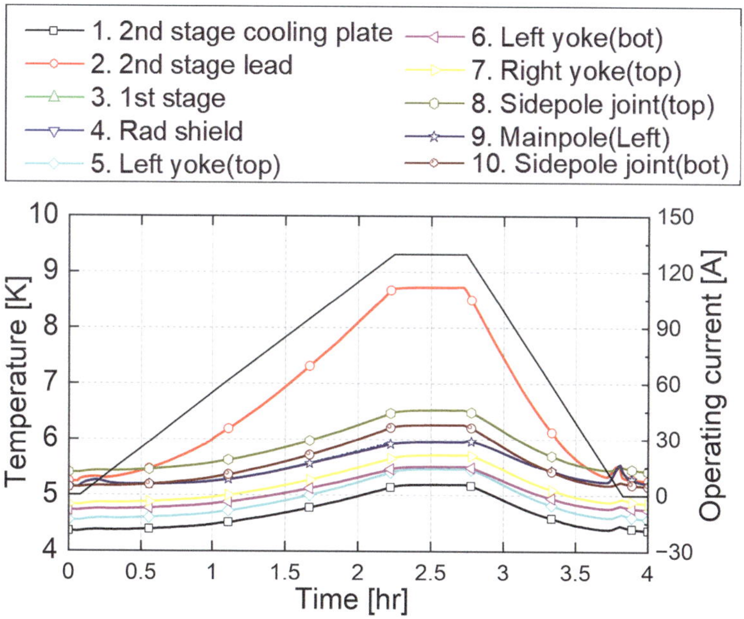

Fig. 6.2 Measured temperatures during the ramping of the magnet in conduction cooling operation

heat load by powering the magnet may be 1.1 W estimated from the cooling capacity curve of the cryocooler. The largest temperature rise happened at the second stage lead, this is mainly because resistive current lead is connected between the magnet and HTS current lead.

6.2.2 Voltage Measurement Results

Figure 6.3a shows the measured magnet terminal. From the results of analyzing the delayed voltage signal with the lumped circuit model in Sect. 3.4, the contact resistivity is calculated by 1,050 $\mu\Omega\cdot cm^2$. There is a reasonable agreement between the measured and simulated magnet voltage. Noise signal may be found at around $t = 1$ h, 2.1 h. These signals may have originated from the movement of the magnet by the increased Lorentz force.

Figure 6.3 shows the measured joint voltages. Ideally, the joint resistance is linearly proportional to the operating current, however, the resistance of joint 5 dramatically increased when the operating current is higher than 50 A. The cause of increased resistance might be the degradation of joints during the manufacturing

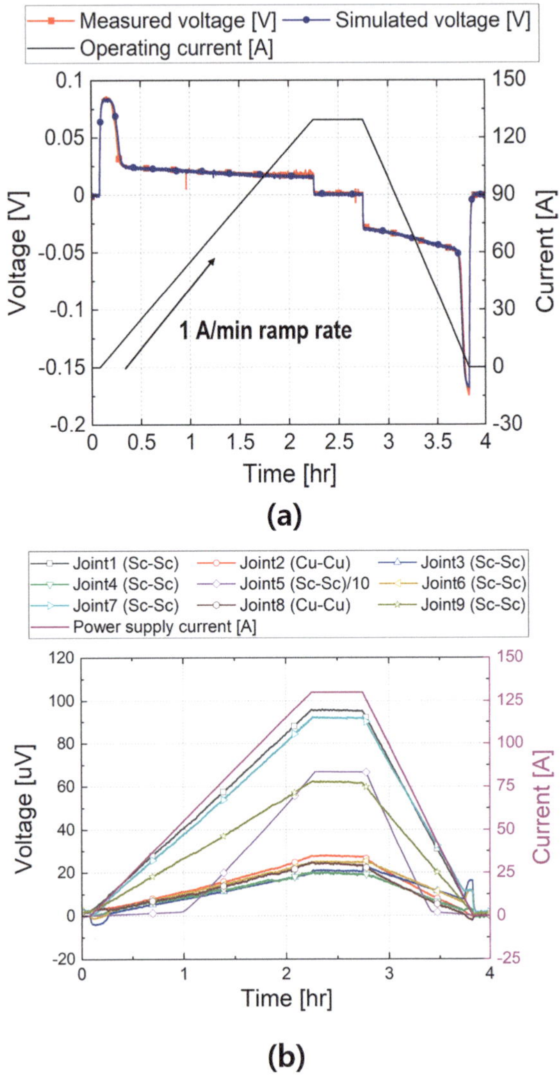

Fig. 6.3 Measured voltages during the ramping of the magnet: **a** terminal voltage; **b** joint voltage

process. Table 6.2 shows the measured resistances of every joint, where the ranges of the surface resistivity of the joint range from 300 to 1,300 nΩ·cm^2. It is known that the resistivity of the joint varies by the conductor spool. The total resistance of the joint is around 8.1 μΩ including the degraded joint 5, and the heat generated from these joint resistances is affordable in this cooling system.

Table 6.2 Measured joint resistivity

Joint number	1	2	3	4	5	6	7	8	9
R [nΩ]	714	181	134	157	312	176	662	160	454
R_{ct} [nΩcm^2]	585	362	268	314	331	440	1332	320	372

6.2.3 *Field Measurement Results*

Figure 6.4 shows the measured magnetic field when the magnet is charged. The slope of the field decreases as the operating current increases due to the saturation of iron. The dramatic change of slope occurs around 2 T, which is near the saturation field of iron. The estimated magnetic field with an operating current of 130 A is 5.11 T, but the measured field is 5.02 T. The difference between the simulation and measurement is due to the engineering uncertainty and screening current-induced field. The magnetic gap is designed to be 12 mm, however, the measured magnetic gap is 12.4 mm due to the engineering error. The increased magnetic gap is one of the causes of the reduction in the center field. The effect of screening current-induced field is discussed in the following section.

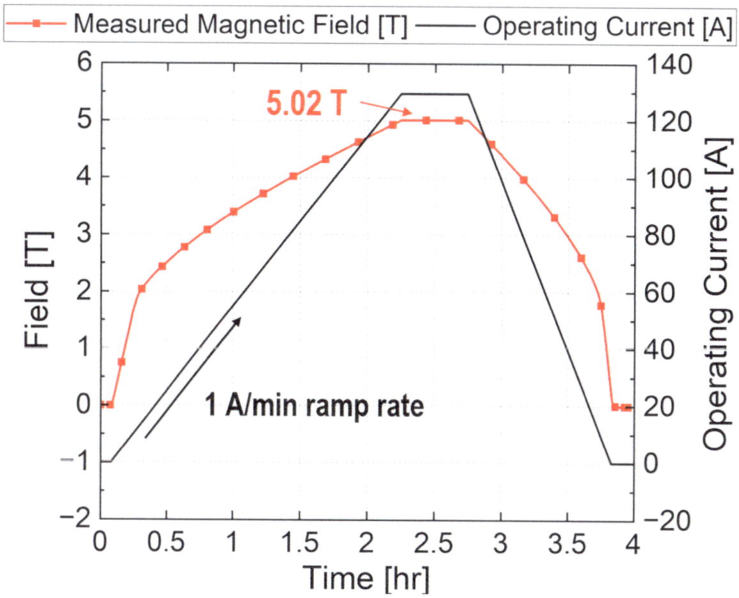

Fig. 6.4 Measured magnetic field of the WLS magnet in conduction cooling operation

6.3 Analysis on Effects of Screening Current-Induced Field

6.3.1 Modeling of Screening Current-Induced Field

6.3.1.1 H-Formulation with Linear Permeability

Basically, the Faraday law of induction is the governing equation of H-formulation-based FEM simulation method.

$$\begin{aligned} \nabla \times E &= \frac{\partial B}{\partial t}, \\ \nabla \times H &= J, \\ E &= \rho J = \rho \nabla \times H. \end{aligned} \tag{6.1}$$

In the two-dimensional space, the vector of the H and E may be simplified to $[H_x, H_y, 0]$ and $[0, 0, E_z]$. If the permeability of material is constant, Eq. (6.1) can be represented by these variables as follows:

$$\begin{aligned} \frac{\partial E_z}{\partial x} &= \mu_0 \frac{\partial H_y}{\partial t}, \\ -\frac{\partial E_z}{\partial y} &= \mu_0 \frac{\partial H_x}{\partial t}, \end{aligned} \tag{6.2}$$

where μ_0 is the permeability of the material. The property of the superconductor is considered in the E-J constitutive law. The resistivity of a superconductor is described by the well-known power law model as shown in Fig. 6.5.

$$\rho = \frac{E_c}{J_c(B,\theta)} \left| \frac{J}{J_c(B,\theta)} \right|^{(n-1)}. \tag{6.3}$$

Therefore, Eq. (6.1) may be represented by the general form of partial derivatives as follows:

$$\begin{aligned} &e_a \frac{\partial^2 u}{\partial t^2} + d_a \frac{\partial u}{\partial t} + \nabla \cdot \Gamma = f, \\ &u = \begin{bmatrix} H_x \\ H_y \end{bmatrix}, \\ &\Gamma = \begin{bmatrix} 0 & -E_z \\ E_z & 0 \end{bmatrix}, \\ &e_a = 0, \\ &d_a = \begin{bmatrix} \mu & 0 \\ 0 & \mu \end{bmatrix}, \\ &f = 0. \end{aligned} \tag{6.4}$$

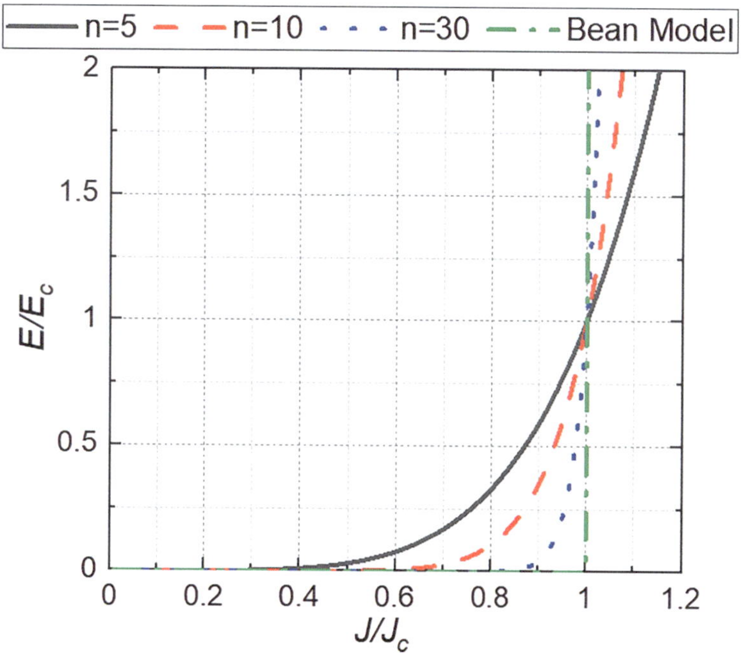

Fig. 6.5 Nonlinear resistance of superconductor modeled by power law

The screening current density is represented by

$$J_z = -\frac{\partial H_x}{\partial y} + \frac{\partial H_y}{\partial x}. \tag{6.5}$$

6.3.1.2 Modified H-Formulation Considering Nonlinear BH Curve of Iron Core [57]

Derivation of H-formulation with nonlinear BH curve considered starts from Eq. (6.1).

$$\begin{aligned}
\nabla \times E &= \frac{\partial E_z}{\partial y}\hat{x} - \frac{\partial E_z}{\partial x}\hat{y},\\
-\frac{\partial B}{\partial t} &= -\mu_0 \frac{\partial}{\partial t}(\mu_r H_x \hat{x} + \mu_r H_y \hat{y})\\
&= -\mu_0 \left[\frac{\partial H_x}{\partial t}\frac{\partial}{\partial H_x}(\mu_r H_x \hat{x}) + \frac{\partial H_y}{\partial t}\frac{\partial}{\partial H_y}(\mu_r H_x \hat{x}) \right.\\
&\quad \left. + \frac{\partial H_x}{\partial t}\frac{\partial}{\partial H_x}(\mu_r H_y \hat{y}) + \frac{\partial H_y}{\partial t}\frac{\partial}{\partial H_y}(\mu_r H_y \hat{y}) \right]\\
&= -\mu_0 \left[\left\{ \frac{\partial H_x}{\partial t}\left(\mu_r + \frac{\partial \mu_r}{\partial H_x}H_x\right) + \frac{\partial H_y}{\partial t}\frac{\partial \mu_r}{\partial H_y}H_x \right\}\hat{x} \right.\\
&\quad \left. + \left\{ \frac{\partial H_y}{\partial t}\left(\mu_r + \frac{\partial \mu_r}{\partial H_y}H_y\right) + \frac{\partial H_x}{\partial t}\frac{\partial \mu_r}{\partial H_x}H_y \right\}\hat{y} \right].
\end{aligned} \tag{6.6}$$

μ_r is defined by

$$\mu_r \equiv \frac{f(H)}{H}, \tag{6.7}$$

where $f(H)$ represents the BH curve of material, where nonlinear curve for iron, while linear curve with a slope of vacuum permeability for air. The derivative of μ_r is obtained by

$$\begin{aligned}
\frac{\partial \mu_r}{\partial H} &= -\frac{B}{H^2} + \frac{1}{H}\frac{\partial B}{\partial H} = \frac{1}{H^2}\left[-B + H\frac{\partial B}{\partial H} \right],\\
\frac{\partial \mu_r}{\partial H_x} &= \frac{\partial H}{\partial H_x}\frac{\partial \mu_r}{\partial H} = \frac{H_x}{H}\frac{\partial \mu_r}{\partial H},\\
\frac{\partial \mu_r}{\partial H_y} &= \frac{\partial H}{\partial H_y}\frac{\partial \mu_r}{\partial H} = \frac{H_y}{H}\frac{\partial \mu_r}{\partial H}.
\end{aligned} \tag{6.8}$$

Finally, H-formulation considering the nonlinear property of iron core in two-dimensional Cartesian coordinates can be formulated by

$$\frac{\partial E_z}{\partial x} = \Lambda H_x H_y \frac{\partial H_x}{\partial t} + \left[\frac{f(H)}{H} + \Lambda H_y^2 \right]\frac{\partial H_y}{\partial t}, \tag{6.9a}$$

$$-\frac{\partial E_z}{\partial y} = \left[\frac{f(H)}{H} + \Lambda H_x^2 \right]\frac{\partial H_x}{\partial t} + \Lambda H_x H_y \frac{\partial H_y}{\partial t}, \tag{6.9b}$$

$$\Lambda = \frac{1}{H^3}\left[-f(H) + H\frac{\partial f(H)}{\partial H} \right], \tag{6.9c}$$

where H_x, H_y, H, E_z, and t are x-axis magnetic field, y-axis magnetic field, the magnitude of magnetic field, electric field, and time[118–120]. The denominator is replaced by an infinitesimally small value to avoid zero denominator, when H is zero. E-J constitutive law for modeling resistance of superconductor used in this study is given by

$$E_z = E_c\left(\frac{|J|}{J_c(|B|,\theta)}\right)^n, \tag{6.10}$$

where $E_c, n, |B|, \theta, l_f$, and J are electric field criteria of 1 μV/cm, n-value, the magnitude of the magnetic field, magnetic field angle, ratio of critical current compared to the reference sample, and current density. Measured field and angle-dependent critical current of REBCO conductor, $I_c(|B|, \theta)$, is referred to [65]. Besides, numerical techniques such as linear edge element and domain homogenization technique are adopted [118]. Numerical methods are implemented by a commercial finite element analysis software, comsol multiphysics.

6.3.1.3 FEM Modeling of Screening Current Analysis of WLS Magnet

In reality, current in REBCO conductor flows in a very thin layer since the thickness of REBCO is about 1 μm. However, modeling a very thin structure may increase the complexity of meshing the geometry. In addition, magnet consists of thousands of turns, and modeling of all the turns may increase the computational load. To address these issues, we have adopted the domain homogenization method.

$$\frac{I_t}{D} = \int J_z dy, \tag{6.11}$$

where I_t is the transport current and other notations are represented in Fig. 6.6. Figure 6.7 shows the geometry of the WLS magnet to calculate screening current distributions within the winding. Symmetry of the magnet structure enabled modeling quarter geometry of the magnet with the given boundary conditions:

$$H_x = 0. \tag{6.12}$$

Here, we neglected the screening current distributions of the side pole.

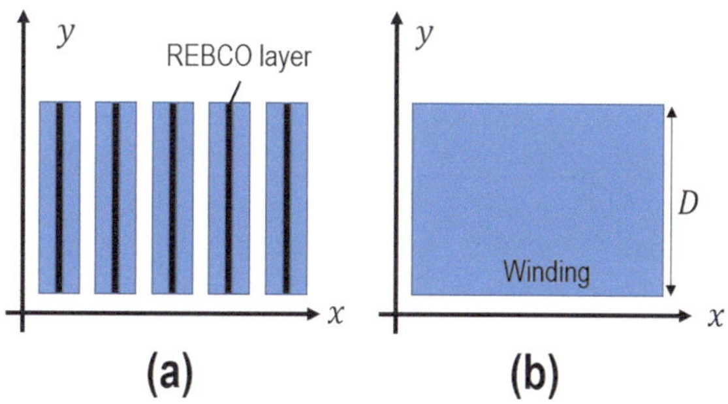

Fig. 6.6 Schematic drawing to explain domain homogenization technique

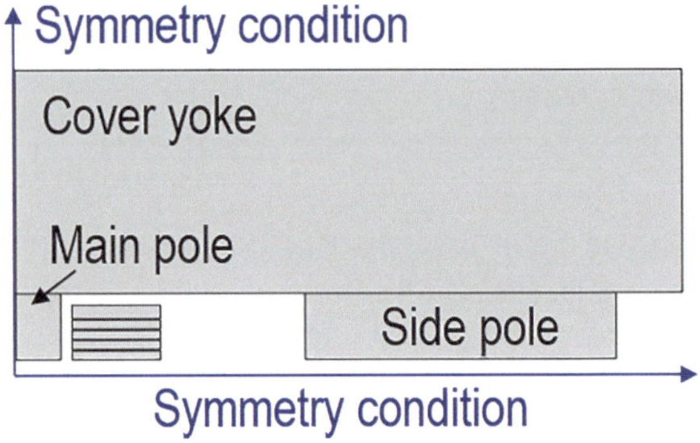

Fig. 6.7 FEM geometry for screening current analysis modeling

6.3.1.4 Segmentation Method to Calculate 3D Field Distributions

Nowadays, there are versatile tools to simulate 3D screening current distributions of HTS magnets. However, the major challenge of the 3D screening current analysis is the complexity of modeling. Recent studies on the 3D screening current distribution analysis are limited to studies on a stack of tapes [121, 122], bulk[123], and model coil [124]. Modeling of magnet scale 3D screening current simulation has not been performed yet. Here, we adopted the segmentation method to efficiently calculate 3D field distributions, while the screening current distributions are calculated from the 2D analysis model.

The segmentation method was originally proposed to accurately calculate field homogeneity of HTS magnets considering screening currents [79]. The key idea of the segmentation method is to divide the conductor flowing region into several segments and calculate the equivalent loop currents using an equivalent current transformation technique. Figure 6.8 shows the schematic diagram of a segmented domain comprised of $2 \times p \times q$ segments [79]. Each triangular segment has three nodes denoted by a, b, and c. The equivalent (I_{equiv}) current flowing at the center of the domain can be represented by the following equation:

$$I_{equiv} = J_a S_a + J_b S_a + J_c S_c. \tag{6.13}$$

In the following analysis, we divided the current flow region of the main pole coils into 25×40 segments, in total of 1,000 segments. From the obtained equivalent current loops, 3D FEM simulation is performed to calculate 3D magnet field distributions along the beam axis.

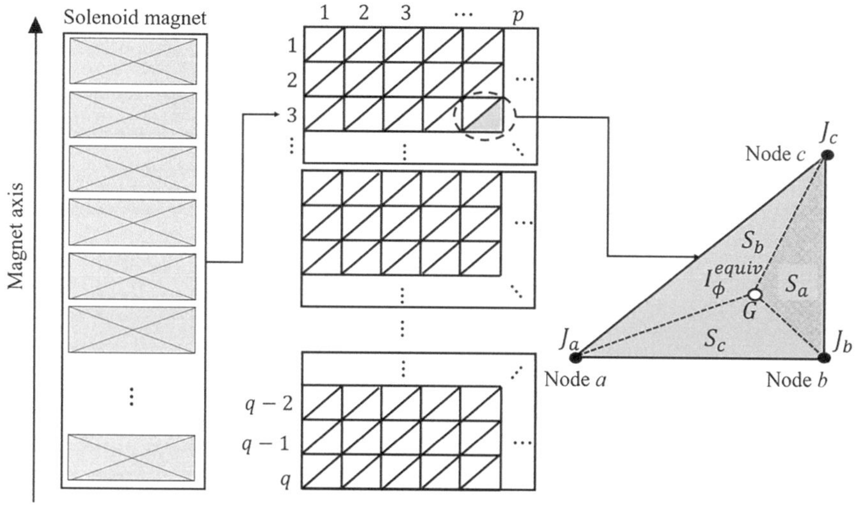

Fig. 6.8 Schematic of $2 \times p \times q$ segmented current flowing region [79]

6.3.2 Screening Current Analysis Results

Figure 6.9 shows the calculated screening current considered total current distributions of the main pole winding at the first charging-discharging of the magnet. Since the *c*-axis field is large at the middle and the uppermost coils, the degree of screening current penetration becomes larger. After the full discharging of the magnet, there are residual current distributions within the winding. Due to this residual current distribution, there exists a residual magnetic field. Although the residual current, the total integrated current density is zero when the magnet is fully discharged.

Figure 6.10 shows the calculated current distributions at the second charging-discharging of the magnet. As the current is charged, screening current flows from the edge of the coil, and the reversed coil current returns to its original state. The calculated current distributions are extracted to the 3D configuration and 3D magnetic field simulation is performed to calculate the screening current-induced field. Figure 6.11 shows the calculated magnetic field considering uniform current density and the screening current.

6.3.2.1 Comparison of Simulated and Measured Field During Ramping the Magnet

Figure 6.12 shows the magnetic field hysteresis of the WLS magnet depending on the operating current. Hysteresis at the second charging and discharging is only presented. *y*-axis of the figure shows the magnetic field difference with respect to the one calculated by uniform current density. The solid line shows the measured

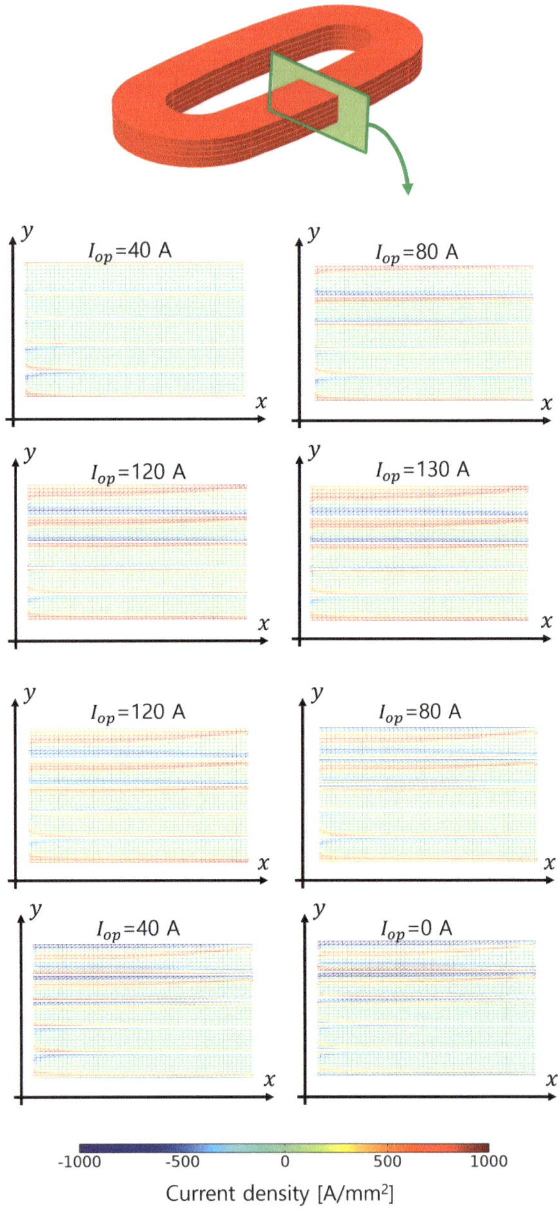

Fig. 6.9 Total current distributions of the main pole winding at the first charging-discharging sequence

Fig. 6.10 Total current distributions of the main pole winding at the second charging sequence

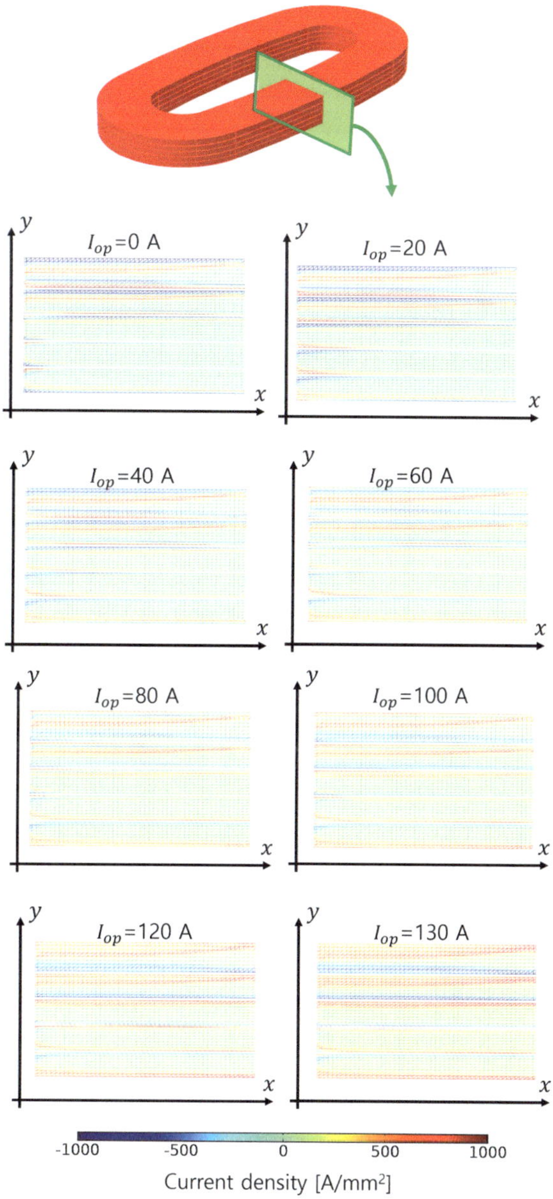

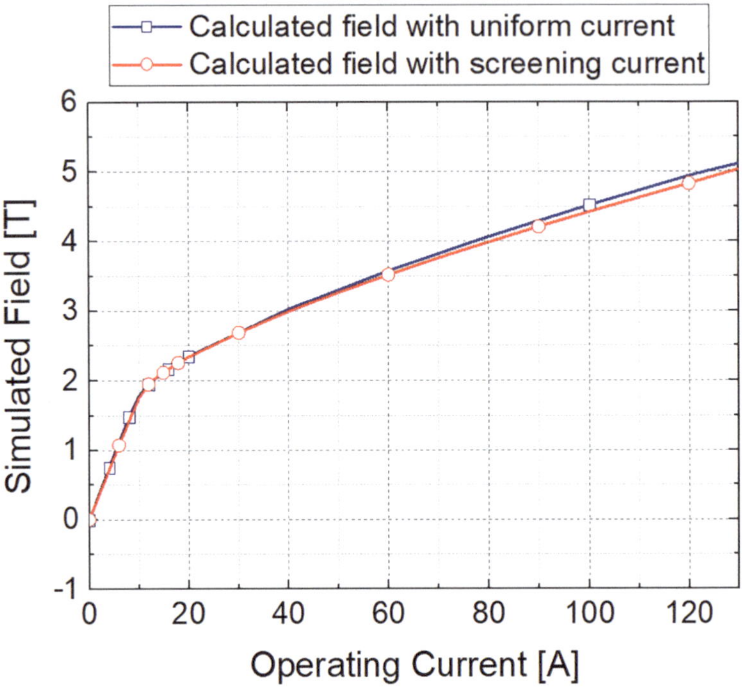

Fig. 6.11 Simulated magnetic field with and without considering the screening current

field hysteresis, while the dashed line shows the field hysteresis calculated by the screening current-induced field. There exists obvious magnetic field hysteresis during the charging and discharging of the magnet. The screening current-induced field reasonably explains this difference.

6.3.2.2 Error Analysis Between the Measurement and Simulation

Figure 6.13 shows the difference in simulated and measured magnetic fields at different operating currents. Screening current considered magnetic field is denoted by B_{scf}, while the one with uniform current density is B_{unf}. At the full operating current of 130 A, the error between the measurement and B_{unf} is about 0.09 T, while it is marginal with B_{scf}. B_{scf} better explains the measured magnetic field at a larger operating current region, however, some errors exist in low-field regions. Here, we discuss two possible sources for the errors in simulation.

- **Measurement error**: Resolution (r_{DAQ}) of data acquisition (DAQ) system can be defined by

$$r_{DAQ} = V_{lim}/A/2^b, \tag{6.14}$$

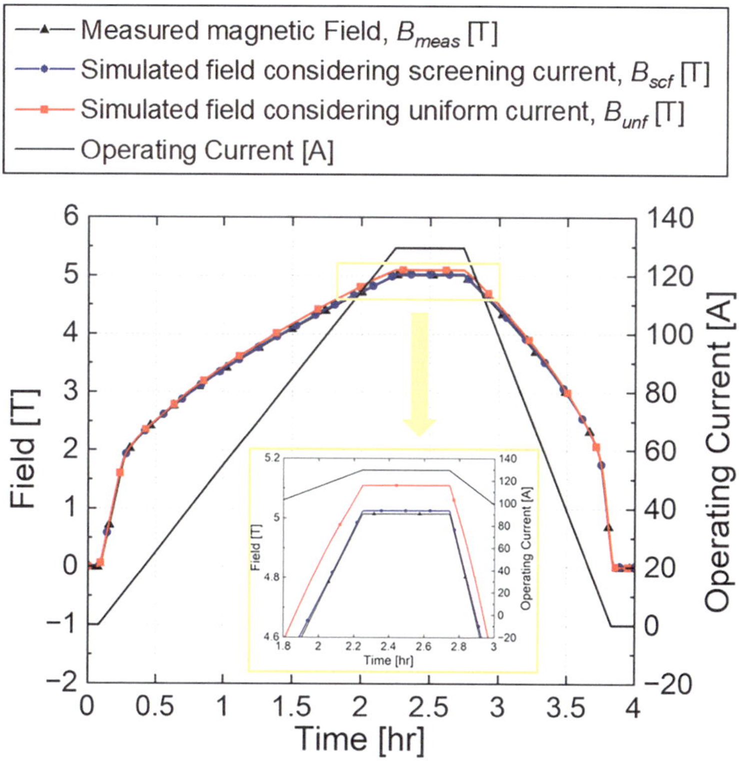

Fig. 6.12 Comparison of the simulated and measured magnetic field when ramping the WLS magnet

where V_{lim}, A, and b are voltage range limit, gain, and bit number of DAQ system. The voltage range limit of SCXI 1125 is 10 V, and bit is 16. The gain of 40 is determined for the Hall sensor voltage measurement. Finally, the DAQ resolution is calculated by 0.4 mT considering the sensitivity of the Hall sensor.

Another measurement error comes from the nonlinearity error of Hall sensor. This phenomenon is known as the Shubnikov-de Haas effect [125]. At a larger current level, the cause of field error comes from the change of linearity of the Hall sensor at cryogenic temperature. Duc to thc galvanomagnetic oscillations, the conductivity of material changes under high magnetic field at cryogenic temperature. This results in a change in the linear sensitivity of the cryogenic Hall sensor. Figure 6.14 shows a change in Hall sensor sensitivity depending on the magnetic field which is measured at the liquid helium temperature.

- **Simulation error**: One of the possible explanations for the error might be uncertainties on the magnetic properties of iron. We used the nonlinear BH curve measured at room temperature as shown in Fig. 3.4. The relative permeability of iron

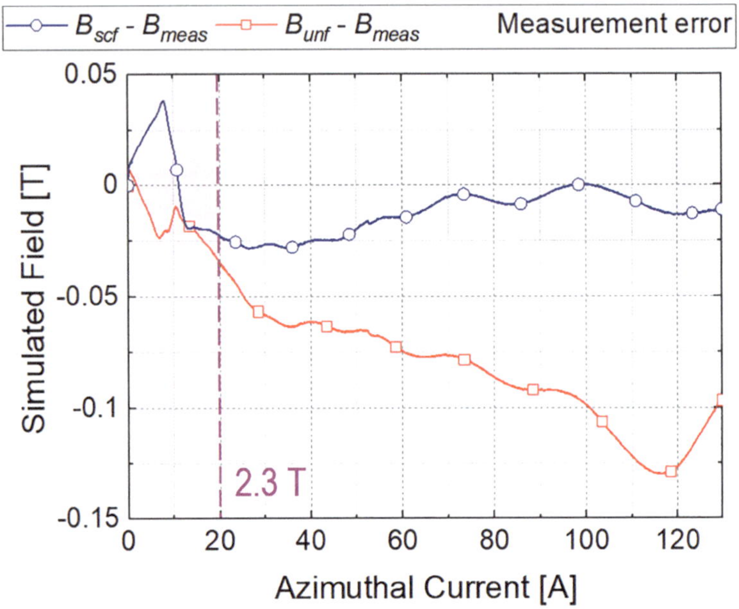

Fig. 6.13 Error of simulated magnetic field with screening current (B_{scf}) and without screening current (B_{unf}) compared with the measurement (B_{meas})

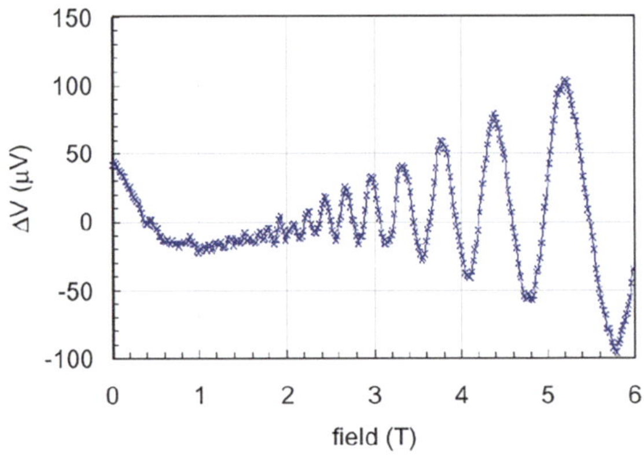

Fig. 6.14 Example of measured Hall sensor voltage variation depending on the magnetic field at 4.2 K [125]

becomes lower at the lower temperature as shown in Fig. 6.15 [126]. Domain walls are less mobile at lower temperatures, thus a larger energy or larger magnetic field is required to form a single domain. Therefore, the relative permeability decreases at lower temperatures.

Two types of iron yoke are considered for this study, low-steel carbon and permendur. Several studies reported cryogenic BH curve measurement results of low-steel carbon[126–129]. However, the results for the permendur iron are rarely reported. BH curve change of permendur pole is rather significant compared to the low-steel

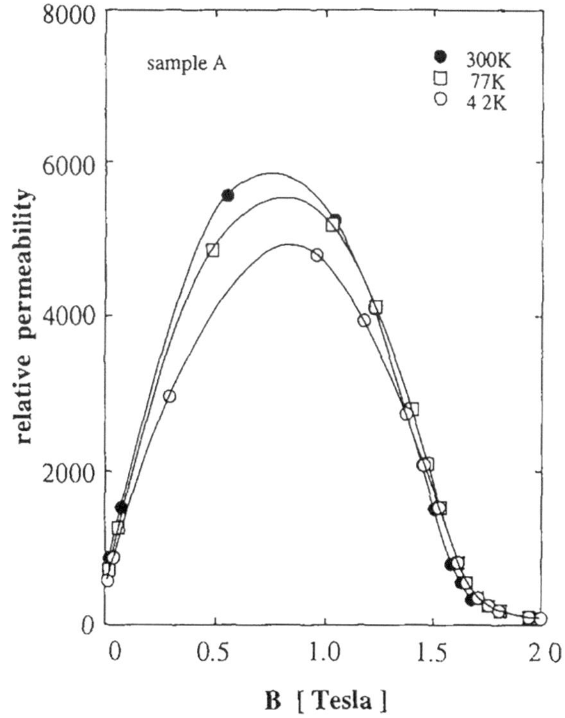

Fig. 6.15 Change of relative permeability of iron in low temperature [126]

carbon since the permendur pole is placed near the magnetic gap. Characterization of the permendur iron at the cryogenic temperature may be further required to more precisely simulate the field distributions.

6.3.3 Beam Orbit Distortion by Screening Current-Induced Field

6.3.3.1 Screening Current-Induced Field Along the Beam Axis

So far, we verified our screening current analysis model by comparing it with the measured magnetic field. From now on we discuss the effect of screening current-induced field during the beam operation. Figure 6.16 shows the screening current-induced field along the beam axis. The total change of field integral is calculated by 4.8×10^{-3} T·m. Estimated kick from the screening current-induced kick is 0.5 mrad. One of the requirements for the insertion device operation is the zero field integral since the beam orbit should be maintained after they pass the insertion devices.

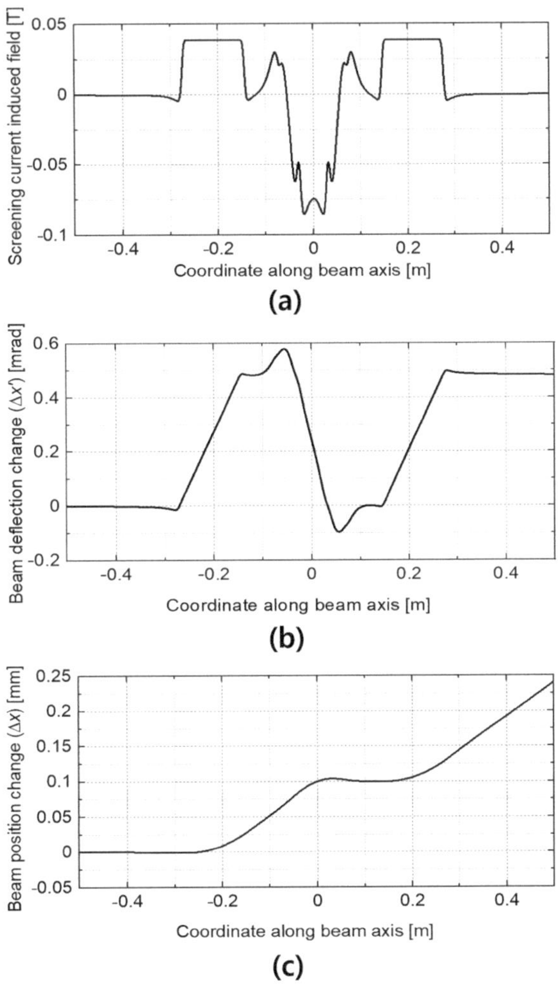

Fig. 6.16 Screening current-induced effects: **a** field change; **b** beam deflection change; **c** beam position change

6.3.3.2 Mitigation of Screening Current-Induced Field Integral by Multiple Power Supply Operation

In the future operation of the WLS magnet, it is possible to operate the magnet by using two power supplies. The side pole coils are connected in series, while the main pole is connected to the individual power supply. Therefore, the operating current of each pole can be controlled individually. Figure 6.17 shows the magnetic field change along the beam axis with an additional operating current of 1 A for side pole coils. The total integral change by the additional side pole current of 1 A is calculated to be

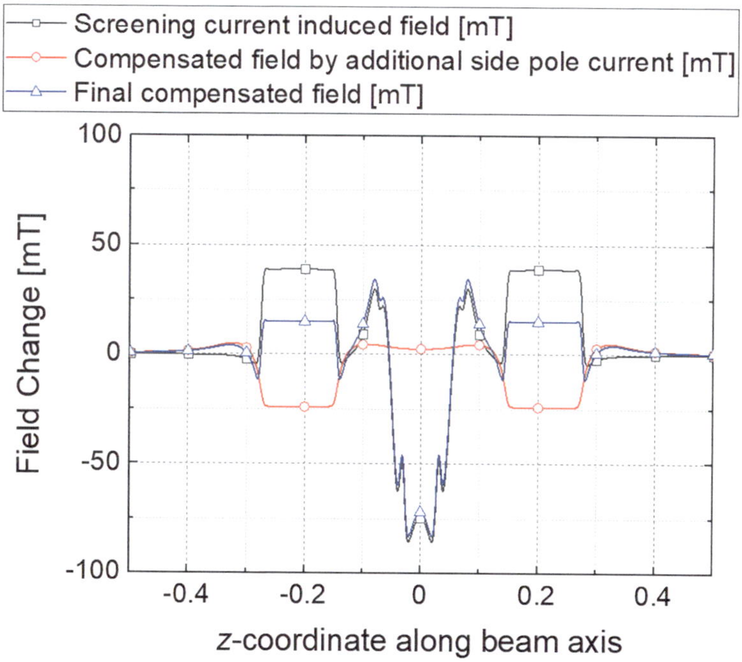

Fig. 6.17 Final compensated field considering screening current-induced field and side pole current

0.88×10^{-4} T·m. With an additional side pole current of −5.5 A, it is anticipated to keep zero field integral even considering the screening current-induced field. Finally, the optimal power supply currents for zero field integral operation are 124.5 A, and 130 A, respectively, for side pole and main pole.

6.4 Future Plans

- **Spatial field measurement**: One of the limitations of this study is that the magnetic field is measured locally at the selected points. Additionally, measurement uncertainties came from the use of uncalibrated cryogenic Hall sensors. There are limited facilities that can precisely measure spatial field distributions in cryogenic and ultra-vacuum conditions. Further studies to precisely evaluate the time stability and spatial homogeneity of the magnetic field may be required with the spatial and temporal measurement of the magnetic field.

- **Accelerator installation**: Only one thermal cycle and two charging-discharging sequences have been investigated for the developed magnet. The magnet stably generated its full field of 5 T for 30 min. No sign of voltage and temperature increase has been observed, but no further tests were done for the developed magnet. The long-term stability of the magnet should be further investigated by installing and operating the magnet in the real accelerator. The developed magnet is currently under preparation for installing the magnet in the actual accelerator.

Chapter 7
Conclusion

This paper presents the design, fabrication, and test results of the first 5 T wavelength shifter(WLS) magnet with high-temperature superconductor(HTS) incorporated. Key questions of the study are the applicability of HTS magnet to the accelerator system, and problems regarding these questions are the manufacturability of HTS accelerator magnet and the effect of screening current induced field during HTS magnet operation. Key challenges are the (1) Lorentz force-induced high mechanical stress, (2) undefined design and manufacturing methods of HTS accelerator magnet, and (3) 3D screening current-induced field analysis model with nonlinear iron characteristics considered.

To solve the difficulties of HTS magnet protection and the high operation cost of liquid helium cooling, we suggested conduction-cooled liquid helium-free metal insulated (MI) HTS WLS magnet for accelerator-based light source. First, we performed the electromagnetic design of the WLS magnet considering the field requirement as well as the critical current property of the HTS conductor. With the parameter sweep method to find optimal multi-width configuration, the magnetic is designed to maximize its main pole field with the limited HTS conductor usage. After that, the final 3D configuration of the magnet is determined to minimize the field integral. Next, we suggested a straight-section reinforcement structure utilizing the existing cooling channel. We applied the new boundary condition called "spring foundation" for the mechanical analysis model to physically model the bent of the fastened bolt on the cooling channel. We designed the size of the fastening bolt considering the affordable mechanical limit as well as maximizing the thermal contact area. After that, we performed a thermal analysis of the magnet considering all the possible heat loads and conduction cooling operation. We confirmed that the maximum temperature of the magnet is kept under 20 K during the magnet operation. Finally, the charging characteristics of the MI HTS magnet are analyzed using the lumped circuit model. To charge and discharge the magnet within a day, we decided to select MI options instead of the bare no-insulation winding.

J. Park, *5 T High Temperature Superconductor 3 Pole Wavelength Shifter Magnet for Accelerator-based Light Sources*, Springer Theses,
https://doi.org/10.1007/978-981-96-8679-7_7

Next, we defined the core manufacturing technologies for the fabrication of the MI HTS WLS magnet including winding, joint, structure, lead, cryogenics, and operation/evaluation. The designed magnet is implemented by combination use of four single pancake coils for the side pole, while four double pancake coils and two single pancake coils for the main pole magnet. We identified the winding dimension difference in the straight section of the racetrack magnet. We explained the difference in the winding dimension with the suggested winding dimension analysis model as well as the compressible characteristics of dry winding. Electrical connection between the identical poles is performed by the bridge joint, while an HTS busbar connection with low melting temperature solder has been implemented. For the structural design of the magnet system, we suggested a new preload structure to connect the upper and lower magnet structures. The preload structure design method is suggested to withstand thermal contraction and the deformation by Lorentz force, and as a result, the strength of fastening torque and Belleville washer stack configurations were determined. For the cryogenic design, we implemented liquid nitrogen precooling to reduce the initial cooling time of the magnet. Finally, a measurement system to monitor magnet voltage, magnetic field, and temperature distributions is constructed.

We fabricated a total of 10 pancake coils and the performance of the coils is individually investigated under liquid nitrogen circumference. The critical current with 0.1 μV/cm criteria, joint resistivity, and contact resistances are measured with the overcurrent charging test and fast discharging test. The order of coil stacking is determined to maximize the critical current of the final magnet and installed in the in-house conduction cooling system.

Finally, the fabricated WLS magnet is tested under conduction cooling circumference. It took 34 h to reach the final temperature including the precooling process. The second stage temperature distributions of the magnet range from 4.5 to 5.5 K, so we confirmed that the magnet is fabricated without a loss of thermal conduction path. The magnet is ramped up to the operating current of 130 A and reaches its main pole magnetic field of 5 T. The temperature of the 2nd stage cooling plate is increased by 0.8 K which corresponds to an additional heat load of 1.1 W. To explain the difference between the simulated and the measured magnetic field, we developed iron considered screening current-induced field analysis model. We formulated a nonlinear BH curve considered H-formulation for screening current analysis and extended the current distributions to 3D configuration by using the segmentation method. Both temporal stability and spatial stability of the measured magnetic field have a reasonable agreement with the calculated magnet field with the developed analysis model. Based on the analysis model, we confirmed that the magnetic field integral caused by the screening current can be canceled through multiple power supply operations of the WLS magnet.

Appendix
Liquid Nitrogen Test Results of Individual Pancake Coils

See Figs. A.1, A.2, A.3, A.4, A.5 and A.6.

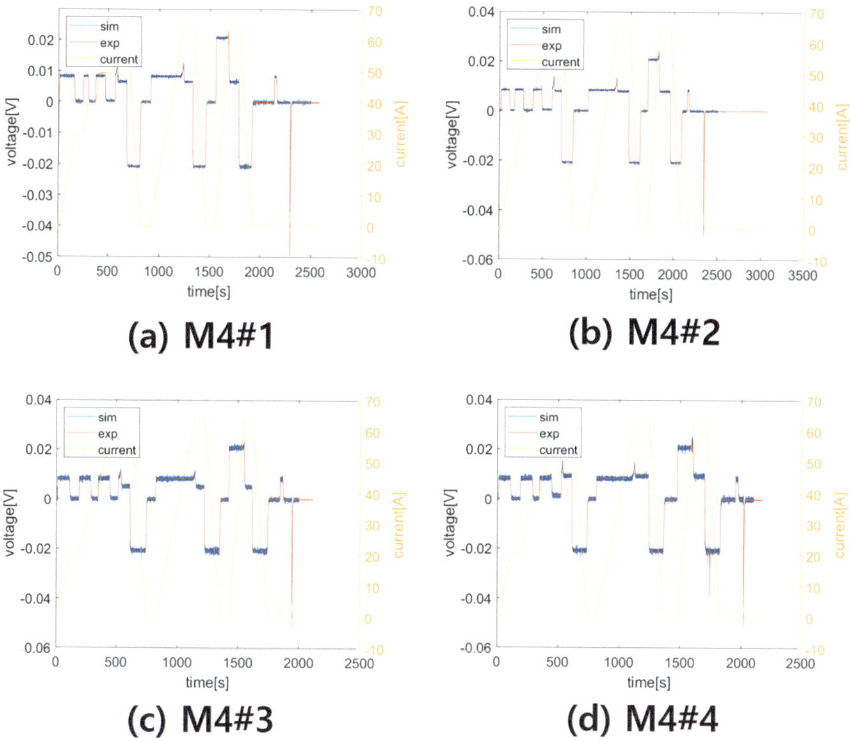

Fig. A.1 Comparison of measured and simulated voltage during the charging-discharging test: **a** M4#1; **b** M4#2; **c** M4#3; **d** M4#4

J. Park, *5 T High Temperature Superconductor 3 Pole Wavelength Shifter Magnet for Accelerator-based Light Sources*, Springer Theses,
https://doi.org/10.1007/978-981-96-8679-7

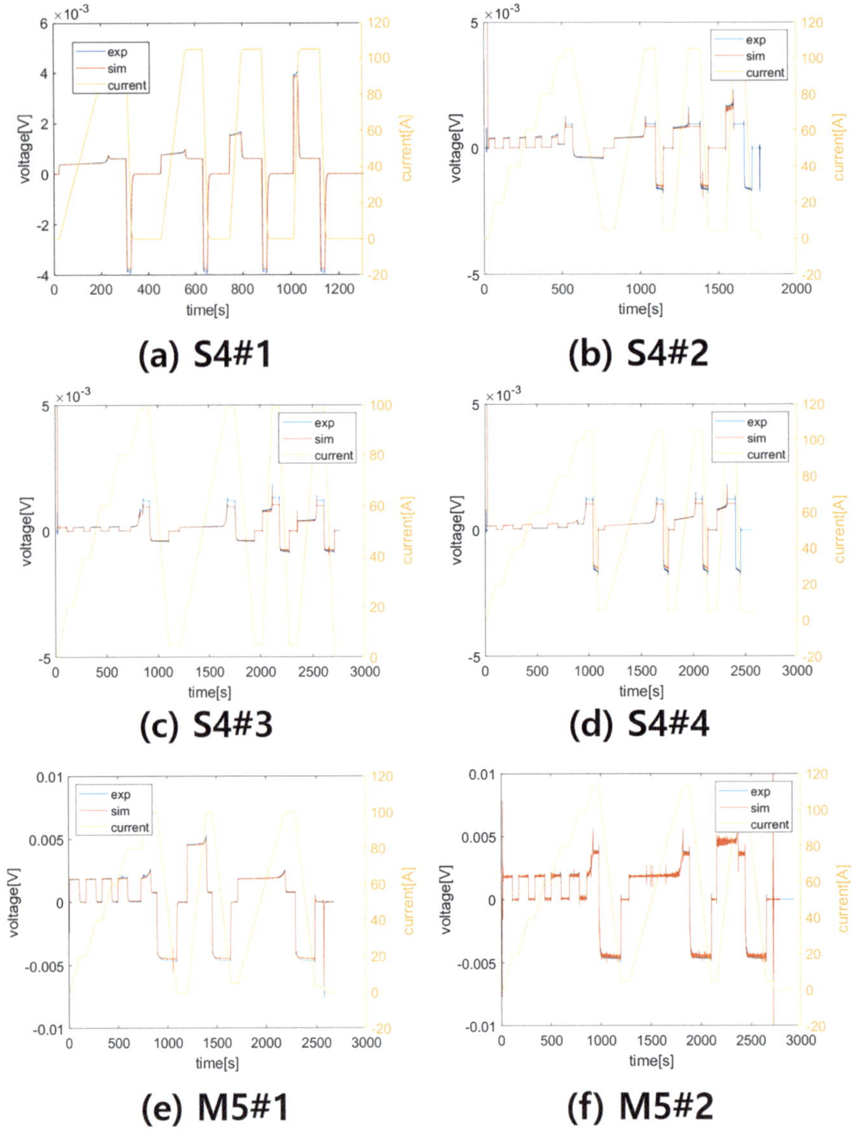

Fig. A.2 Comparison of measured and simulated voltage during charging-discharging test: **a** S4#1; **b** S4#2; **c** S4#3; **d** S4#4; **e** S5#1; **f** S5#2

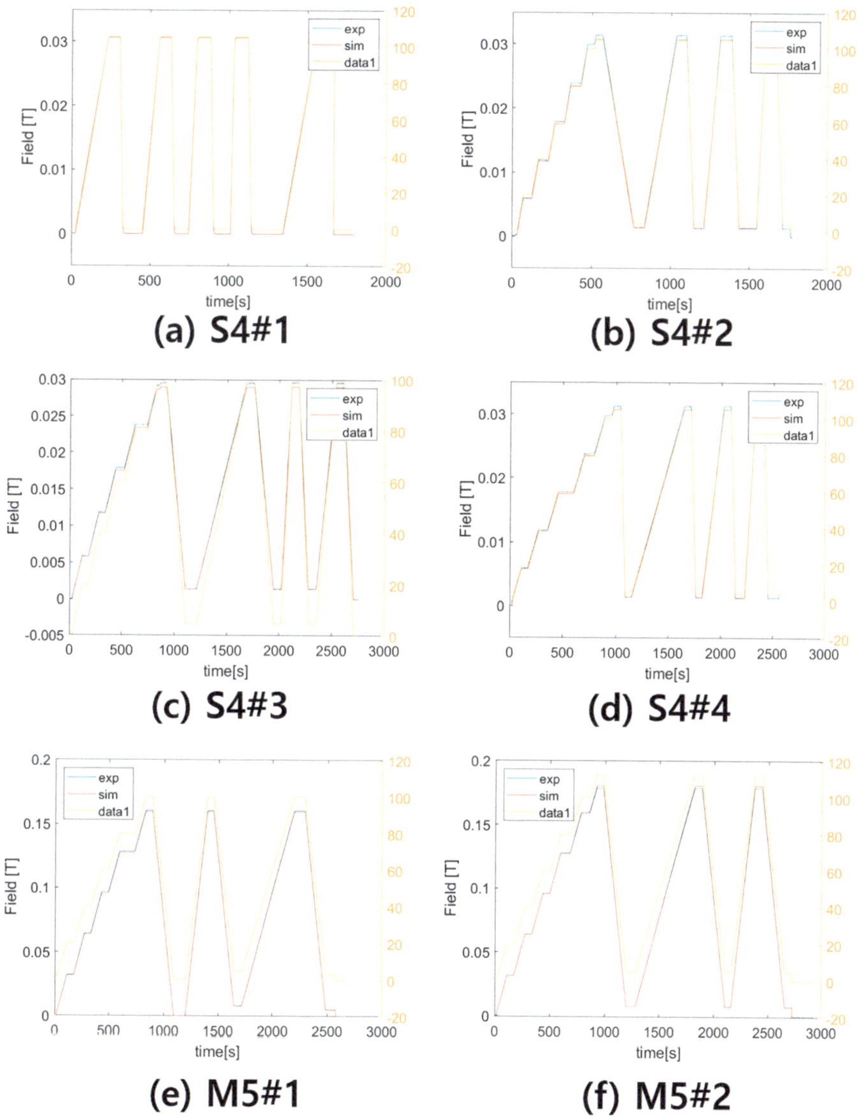

Fig. A.3 Comparison of measured and simulated magnetic field during charging-discharging test: **a** S4#1; **b** S4#2; **c** S4#3; **d** S4#4; **e** S5#1; **f** S5#2

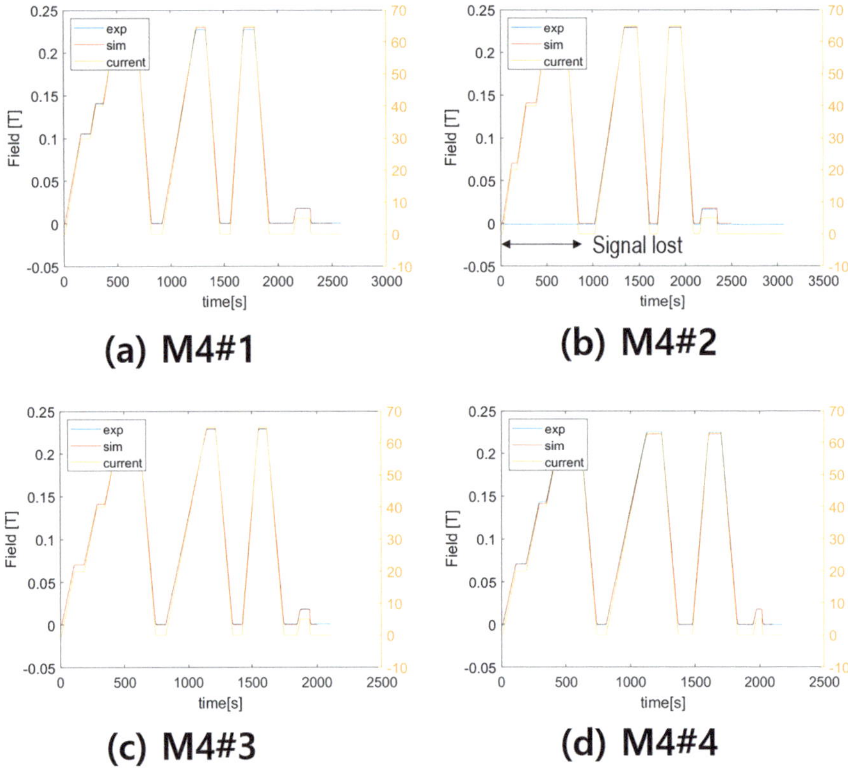

Fig. A.4 Comparison of measured and simulated magnetic field during charging-discharging test: **a** M4#1; **b** M4#2; **c** M4#3; **d** M4#4

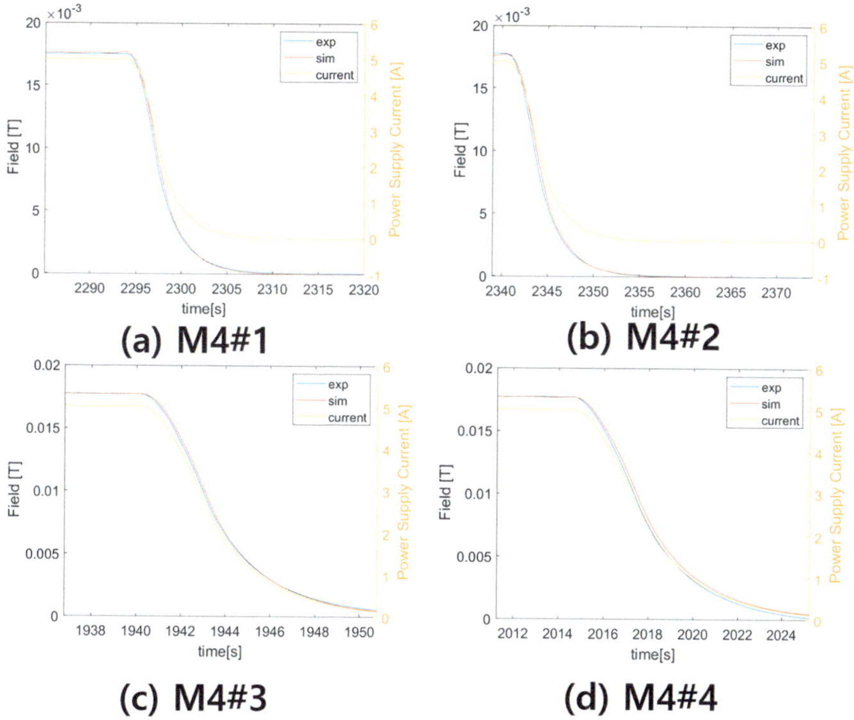

Fig. A.5 Comparison of measured and simulated magnetic field during sudden discharge test: **a** S4#1; **b** S4#2; **c** S4#3; **d** S4#4; **e** S5#1; **f** S5#2

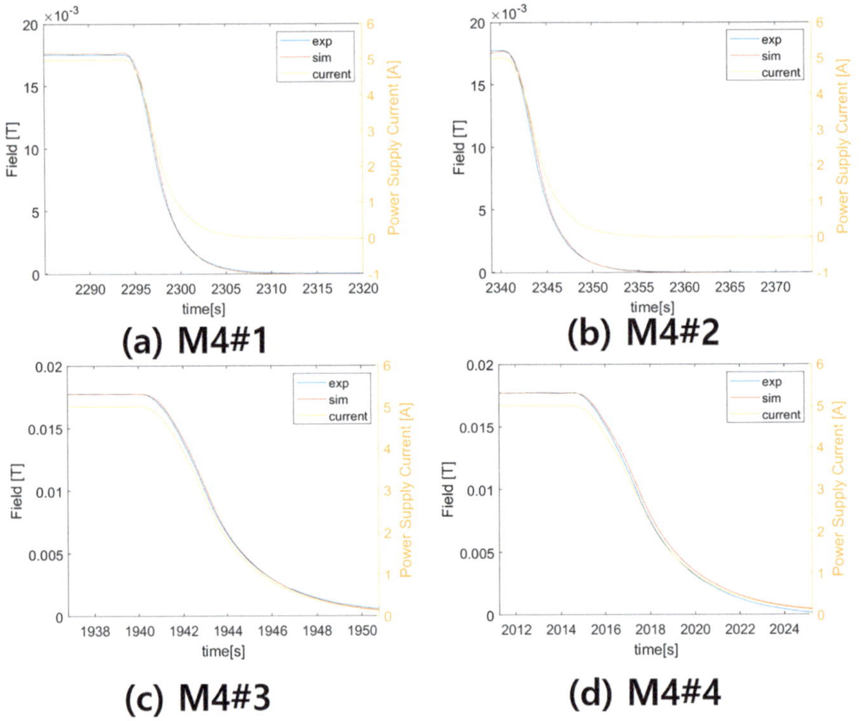

Fig. A.6 Comparison of measured and simulated magnetic field during sudden discharge test: **a** M4#1; **b** M4#2; **c** M4#3; **d** M4#4

초록

본 논문에서는 세계 최초로 고온초전도 기술이 적용된 5 T 전도냉각형 금속절연 고온초전도 파장변환기 자석을 개발하였다. 전도냉각형 금속절연 고온초전도 파장변환기 자석의 설계, 제작, 운전을 포함한 전반적인 개발 과정을 다루는 것인 본 논문의 주요 범위이다. 우선, 전자기, 기계, 열, 운전 특성을 고려한 전도냉각형 금속절연 파장 변환기 자석의 설계 방법을 제안하였다. 다음으로 전도냉각형 금속절연 고온초전도 파장변환기 자석 개발을 위한 핵심 제조 기술을 개발하였다. 또한, 고온초전도 파장변환기 자석의 자기장 운전 특성 예측을 위해 철의 비선형 특성을 고려한 3차원 차폐전류자기장 분석 모델을 개발하였다.

가속기 기반 광원은 입자가속기를 이용하여 높은 선량의 광자를 만들어 내는 장치이다. 가속기 기반 광원에서 높은 자기장의 파장변환기 자석을 활용할 경우 더 넓은 에너지 스펙트럼의 광자를 얻을 수 있으며, 이는 발생한 광자를 더 다양한 실험에 활용할 수 있다는 의미를 갖는다. 고온초전도 기술은 높은 자기장을 생성할 수 있는 가능성으로 인해 많은 연구자들에 의해 고온초전도 가속기 자석에 대한 연구가 진행되고 있다. 하지만, 고자기장에 의한 높은 기계적 응력, 고온초전도 가속기 자석의 제작 기술의 부재, 그리고 차폐전류 유도 자기장에 의한 자기장 왜곡 등과 같은 주요 어려움으로 인해 아직까지 실제 가속기에 적용된 바가 없다. 또한 퀜치에 대한 고온초전도 자석 보호의 어려움과 액체 헬륨의 높은 운전 비용은 고온초전도 자석 운전에 있어 추가적인 어려움을 야기한다. 본 논문에서는 액체 헬륨을 사용하지 않는 전도 냉각형 금속절연 고온초전도 자석을 제안하며, 앞서 언급한 문제들을 해결하기 위한 연구 과정을 논의한다.

본 논문에서는 제안된 전도냉각형 금속절연 고온초전도 파장변환기 자석의 설계, 제작, 운전을 포함하여 자석의 전반적인 개발 과정을 제시한다. 먼저, 전도냉각형 금속절연 고온초전도 파장변환기 자석의 설계방법을 제안하였다. 우선, 자기장 균일도 조건 및 자기장 세기의 극대화를 위해 다중폭기법을 적용한 자석의 3차원 형상 설계를 진행하였다. 다음으로, 레이스트랙형 코일의 직선구간의 응력을 저감하지 위한 새로운 보강 구조를 제안하였으며, 제안한 구조를 고려한 기계해석모델을 개발하였다. 또한, 전도냉각 환경을 고려한 열채널 설계를 통해 자석의 최대 온도를 20 K 이하로 제한하는 설계를 진행하였다. 마지막으로, 집중회로 등가모델을 이용하여 금속절연 기법이 적용된 파장변환기 자석의 충방전 지연효과를 분석하였다.

다음으로 전도냉각형 금속절연 고온초전도 파장변환기 자석 개발을 위한 핵심 제조 기술을 개발하였다. 고온초전도 파장변환기 자석의 권선, 접합, 리드, 구조, 극저온, 작동/평가를 포함한 전반적인 제조 과정을 정의하였다. 권선 기술에 대해 권선 장력에 따른 권선 치수 분석 모델을 개발하고, 구조 기술의 경우에는 새로운 예압 구조를 제안하였다. 최종적으로는 제안한 제조 기술을 활용하여 전도 냉각형 금속절연 고온초전도 파장변환기 자석을 성공적으로 개발하였다. 이후, 고온초전도 파장변환기 자석의 자기장 운전 특성 예측을 위해 철의 비선형 특성을 고려한 3차원 차폐전류자기장 분석 모델을 개발하였다. 비선형 철심 특성이 고려된 H-formulation 기반으로 차폐전류 해석모델을 개발하였으며, segmentation 기법을 이용하여 차폐전류 분포에 의한 자기장 왜곡효과를 분석하였다. 개발한 시뮬레이션 모델은 측정한 자기장과 해석 결과를 비교하여 검증하였다. 마지막으로, 입자 빔 운전에서 차폐전류유도 자기장에 의한 효과를 논의하였으며, 부극의 운전 전류의 조정으로 해당 효과를 상쇄시킬 수 있을 것으로 예상하였다.

개발된 고온초전도 파장변환기 자석은 고온초전도 기술이 적용된 최초의 파장변환기 자석이다. 본 연구의 의의 중 하나는 기존 고온초전도 가속기자석에서 CERN이 달성한 최고 자기장인 4.52 T를 넘어 5 T의 자기장 기록을 달성했다는 것이다. 개발된 설계 및 제조 기술은 미래의 고자기장 고온초전도 가속기 자석 개발에 적용될 것으로 기대된다. 또한, 해당 파장변환기 자석 개발은 방사광가속기 핵심부품을 국산화였다는 의의가 있다. 특히, 파장 변환기 자석을 국내 PLS II 저장링에 자석을 적용하면 100 keV 광자 에너지 범위에서 기존 시스템 대비 1,000배 더 밝은 광자를 생성하여, 가속기 기반 광원의 성능을 획기적으로 향상시킬 수 있을 것으로 기대한다.

References

1. E. D. Santarelli, "Fichier:schéma de principe du synchrotron," https://fr.wikipedia.org/wiki/Fichier:Sch%C3%A9ma_de_principe_du_synchrotron.jpg, 2005, [Online; accessed 11-oct-2023].
2. "Just 40 years ago, the photon factory generated its first synchrotron radiation," https://www.kek.jp/en/topics-en/20220623pf40year/, 2022, [Online; accessed 11-oct-2023].
3. K. Wille, *The physics of particle accelerators: an introduction*. Clarendon Press, 2000.
4. J. H. Hubbell and S. M. Seltzer, "X-ray mass attenuation coefficients," https://www.nist.gov/pml/x-ray-mass-attenuation-coefficients, 2022, [Online; accessed 28-Nov-2023].
5. V. Fernandez, F. Abdala, K. J. Carlson, D. C. Cook, B. S. Rubidge, A. Yates, and P. Tafforeau, "Synchrotron reveals early triassic odd couple: injured amphibian and aestivating therapsid share burrow," *PLoS One*, vol. 8, no. 6, p. e64978, 2013.
6. E. J. Jaeschke, S. Khan, J. R. Schneider, and J. B. Hastings, *Synchrotron light sources and free-electron lasers: accelerator physics, instrumentation and science applications*, 2016.
7. K. J. Carlson, D. Stout, T. Jashashvili, D. J. De Ruiter, P. Tafforeau, K. Carlson, and L. R. Berger, "The endocast of mh1, australopithecus sediba," *Science*, vol. 333, no. 6048, pp. 1402–1407, 2011.
8. J. Brunet, B. Pierrat, J. Adrien, E. Maire, N. Curt, and P. Badel, "A novel method for in vitro 3d imaging of dissecting pressurized arterial segments using x-ray microtomography," *Experimental Mechanics*, vol. 61, pp. 147–157, 2021.
9. S. D. Jacques, M. Di Michiel, A. M. Beale, T. Sochi, M. G. O'Brien, L. Espinosa-Alonso, B. M. Weckhuysen, and P. Barnes, "Dynamic x-ray diffraction computed tomography reveals real-time insight into catalyst active phase evolution," *Angewandte Chemie International Edition*, vol. 50, no. 43, pp. 10 148–10 152, 2011.
10. E. Maire and P. J. Withers, "Quantitative x-ray tomography," *International materials reviews*, vol. 59, no. 1, pp. 1–43, 2014.
11. M. of Science and ICT, https://www.msit.go.kr/bbs/view.do?sCode=user&mId=113&mPid=112&pageIndex=&bbsSeqNo=94&nttSeqNo=3180498&searchOpt=ALL&searchTxt=, 2021, [Online; accessed 20-oct-2023].
12. P. F. Tavares, E. Al-Dmour, Å. Andersson, F. Cullinan, B. N. Jensen, D. Olsson, D. K. Olsson, M. Sjöström, H. Tarawneh, S. Thorin, *et al.*, "Commissioning and first-year operational results of the MAX IV 3 GeV ring," *J. Synchrotron Radiat.*, vol. 25, no. 5, pp. 1291–1316, 2018.

J. Park, *5 T High Temperature Superconductor 3 Pole Wavelength Shifter Magnet for Accelerator-based Light Sources*, Springer Theses,
https://doi.org/10.1007/978-981-96-8679-7

13. A. Rodrigues, C. Rodrigues, F. Arroyo, S. Marques, R. Farias, F. Rodrigues, J. Citadini, O. Bagnato, R. Seraphim, L. Liu, *et al.*, "Sirius status report," *Proc. IPAC'16*, pp. 2811–2814, 2016.
14. P. Raimondi, "Esrf-ebs: The extremely brilliant source project," *Synchrotron Radiation News*, vol. 29, no. 6, pp. 8–15, 2016.
15. C. Steier, A. Allézy, A. Anders, K. Baptiste, E. Buice, K. Chow, G. Cutler, R. Donahue, D. Filippetto, J. Harkins, *et al.*, "Status of the conceptual design of ALS-U," in *9th International Particle Accelerator Conference*, 2018, pp. 4134–4137.
16. A. Streun, T. Garvey, L. Rivkin, V. Schlott, T. Schmidt, P. Willmott, and A. Wrulich, "SLS-2–the upgrade of the Swiss Light Source," *J. Synchrotron Radiat.*, vol. 25, no. 3, pp. 631–641, 2018.
17. M. Borland and A. Blednykh, "The upgrade of the advanced photon source," Brookhaven National Lab.(BNL), Upton, NY (United States), Tech. Rep., 2018.
18. E. Karantzoulis, "Elettra 2.0—The diffraction limited successor of Elettra," *Nucl. Instrum. Methods Phys. Res. A.*, vol. 880, pp. 158–165, 2018.
19. H. Tanaka, T. Ishikawa, S. Goto, S. Takano, T. Watanabe, and M. Yabashi, "SPring-8 upgrade project," *Proc. IPAC2016, Busan, Korea*, pp. 2867–2870, 2016.
20. Y. Jiao, G. Xu, X.-H. Cui, Z. Duan, Y.-Y. Guo, P. He, D.-H. Ji, J.-Y. Li, X.-Y. Li, C. Meng, *et al.*, "The HEPS project," *J. Synchrotron Radiat.*, vol. 25, no. 6, pp. 1611–1618, 2018.
21. C. G. Schroer, I. Agapov, W. Brefeld, R. Brinkmann, Y.-C. Chae, H.-C. Chao, M. Eriksson, J. Keil, X. Nuel Gavaldà, R. Röhlsberger, *et al.*, "PETRA IV: the ultralow-emittance source project at DESY," *J. Synchrotron Radiat.*, vol. 25, no. 5, pp. 1277–1290, 2018.
22. J. Ko, J. Lee, B.-H. Oh, K. Jang, M. Yoon, T. Ha, G. Hahn, D.-E. Kim, and S. Shin, "Comparison of photon beam characteristics of Korea-4GSR with PLS-II," *Nucl. Instrum. Methods Phys. Res. A.*, 2021, Art. no. 165941.
23. Y. Ivanyushenkov, J. Fuerst, Q. Hasse, M. Kasa, I. Kesgin, Y. Shiroyanagi, and E. Gluskin, "Status of the development of superconducting undulators at the advanced photon source," *Synchrotron Radiation News*, vol. 31, no. 3, pp. 29–34, 2018.
24. S. Liu, Y. Ding, and J. Xu, "Development of a short rebco undulator magnet with resistive joints," *IEEE Transactions on Applied Superconductivity*, vol. 29, no. 5, 2019, Art. no. 4100204.
25. S. C. Richter, A. Ballarino, D. Schoerling, T. Nes, A. Bernhard, and A.-S. Müller, "Progress on hts undulator prototype coils for compact fel designs," *IEEE Transactions on Applied Superconductivity*, vol. 32, no. 4, 2022, Art. no. 4100305.
26. K. Zhang, A. Pirotta, X. Liang, S. Hellmann, M. Bartkowiak, T. Schmidt, A. Dennis, M. Ainslie, J. Durrell, and M. Calvi, "Record field in a 10 mm-period bulk high-temperature superconducting undulator," *Superconductor Science and Technology*, vol. 36, no. 5, 2023, Art. no. 05LT01.
27. J. Bahrdt and E. Gluskin, "Cryogenic permanent magnet and superconducting undulators," *Nuclear Instruments and Methods in Physics Research Section A: Accelerators, Spectrometers, Detectors and Associated Equipment*, vol. 907, pp. 149–168, 2018.
28. C. Hwang, B. Wang, B. Wahrer, C. Taylor, C. Chen, T. Juang, F. Lin, J. Jan, C. Chang, H. Chen, *et al.*, "Design, construction, and performance testing of a 6.5 t superconducting wavelength shifter," *IEEE transactions on applied superconductivity*, vol. 17, no. 2, pp. 1229–1234, 2007.
29. E. Wallén, "Commissioning of a 6.4 t superconducting wavelength shifter at max-lab," *Nuclear Instruments and Methods in Physics Research Section A: Accelerators, Spectrometers, Detectors and Associated Equipment*, vol. 495, no. 1, pp. 58–64, 2002.
30. A. Grudiev, V. Djurba, G. Kulipanov, V. Khlestov, N. Mezentsev, S. Ruvinsky, V. Shkaruba, S. Sukhanov, P. Vobly, Y. Koo, *et al.*, "Superconducting 7.5 tesla wiggler for pls," *Nuclear Instruments and Methods in Physics Research Section A: Accelerators, Spectrometers, Detectors and Associated Equipment*, vol. 359, no. 1-2, pp. 101–106, 1995.
31. N. Mezentsev and E. Wallén, "Superconducting wigglers," *Synchrotron radiation news*, vol. 24, no. 3, pp. 3–9, 2011.

32. J. Zbanik, S. Wang, J. Chen, G. DeVries, R. DeMarco, M. Fahmie, A. Geyer, M. Green, J. Harkins, T. Henderson, *et al.*, "Als superbend magnet system," *IEEE transactions on applied superconductivity*, vol. 11, no. 1, pp. 2531–2534, 2001.
33. S. Casalbuoni, N. Glamann, A. Grau, T. Holubek, D. Saez de Jauregui, S. Bauer, C. Boffo, T. Gerhard, M. Turenne, and W. Walter, "Superconducting undulators: from development towards a commercial product," *Synchrotron Radiation News*, vol. 31, no. 3, pp. 24–28, 2018.
34. M. Kasa, E. Anliker, Q. Hasse, H. Hu, I. Kesgin, E. Moog, Y. Shiroyanagi, and Y. Ivanyushenkov, "Design, fabrication, and testing of a 1.9-m-long, 16.5-mm period nbti superconducting undulator for the advanced photon source upgrade," *IEEE Transactions on Applied Superconductivity*, vol. 32, no. 6, p. 4101805, 2022.
35. I. Kesgin, M. Kasa, S. MacDonald, D. Turrioni, E. Barzi, Y. Shiroyanagi, Q. Hasse, Y. Ivanyushenkov, A. Zlobin, and E. Gluskin, "Design, construction, and testing of 0.5-m, 18-mm period nb 3 sn superconducting undulator magnets," *IEEE Transactions on Applied Superconductivity*, vol. 32, no. 6, p. 4100605, 2022.
36. J. Ko, B. Oh, S. Shin, J. Lee, G. Jang, and M. Yoon, "Characteristics of the synchrotron radiation of fourth-generation storage rings," *Journal of the Korean Physical Society*, vol. 77, pp. 363–367, 2020.
37. C. Barth, G. Mondonico, and C. Senatore, "Electro-mechanical properties of rebco coated conductors from various industrial manufacturers at 77 k, self-field and 4.2 k, 19 t," *Superconductor Science and Technology*, vol. 28, no. 4, 2015, Art. no. 045011.
38. H.-S. Shin, M. B. De Leon, and M. A. E. Diaz, "Investigation of the electromechanical behaviors in Cu-stabilized GdBCO coated conductor tapes using high-cycle fatigue tests at 77 K and related fractographic observations," *Superconductor Science and Technology*, vol. 33, no. 2, p. 025012, 2020.
39. H.-S. Shin, M. Velasco, and M. A. Diaz, "A practical stress-based test method for evaluating reversible stress limit for critical current degradation in rare-earth barium copper oxide (rebco) tapes," *Review of Scientific Instruments*, vol. 94, no. 10, 2023.
40. N. C. Allen, L. Chiesa, and M. Takayasu, "Numerical and experimental investigation of the electromechanical behavior of rebco tapes," in *IOP Conference Series: Materials Science and Engineering*, vol. 102, no. 1. IOP Publishing, 2015, p. 012025.
41. T. Watanabe, S. Nagaya, A. Ishiyama, S. Noguchi, H. Ueda, and G. Nishijima, "Mechanical strength evaluation of a yoroi-coil structured non-circular rebco pancake coil in high magnetic field," *IEEE Transactions on Applied Superconductivity*, vol. 32, no. 6, 2022, Art. no. 4600605.
42. S. Nagaya, T. Watanabe, T. Tamada, M. Naruse, N. Kashima, T. Katagiri, N. Hirano, S. Awaji, H. Oguro, and A. Ishiyama, "Development of high strength pancake coil with stress controlling structure by rebco coated conductor," *IEEE transactions on applied superconductivity*, vol. 23, no. 3, 2012, Art. no. 4601204.
43. X. Wang, A. Ishiyama, T. Tsujimura, H. Yamakawa, H. Ueda, T. Watanabe, and S. Nagaya, "Numerical structural analysis on a new stress control structure for high-strength rebco pancake coil," *IEEE transactions on applied superconductivity*, vol. 24, no. 3, 2013, Art. no. 4601605.
44. T. Watanabe, S. Nagaya, N. Hirano, S. Awaji, H. Oguro, Y. Tsuchiya, T. Omura, S. Nimori, T. Shimizu, A. Ishiyama, *et al.*, "Strengthening effect of "yoroi-coil structure" against electromagnetic force," *IEEE Transactions on Applied Superconductivity*, vol. 25, no. 3, 2014, Art. no. 4601204.
45. N. M. I. Center, "Helium statistics and information," https://www.usgs.gov/centers/national-minerals-information-center/helium-statistics-and-information, 2023, [Online; accessed 20-oct-2023].
46. Y. Iwasa, *Case studies in superconducting magnets: design and operational issues*. Springer science & business media, 2009.
47. X. Hu, A. Polyanskii, D. Abraimov, A. Gavrilin, H. Weijers, F. Kametani, J. Jaroszynski, and D. Larbalestier, "Analysis of local burnout in a sub-scale test coil for the 32 t magnet after spontaneous quenches during fast ramping," *Superconductor Science and Technology*, vol. 35, no. 7, 2022, Art. no. 075009.

48. T. Ogata, K. Shirai, A. Ishiyama, H. Ueda, S. Noguchi, T. Watanabe, S. Nagaya, and M. Fukuda, "Numerical evaluation of screening current-induced magnetic field considering variable output energy in rebco coil system of skeleton cyclotron for cancer therapy," *IEEE Transactions on Applied Superconductivity*, vol. 31, no. 5, 2021, Art. no. 4400105.
49. M. C. Ahn, T. Yagai, S. Hahn, R. Ando, J. Bascuñán, and Y. Iwasa, "Spatial and temporal variations of a screening current induced magnetic field in a double-pancake hts insert of an lts/hts nmr magnet," *IEEE transactions on applied superconductivity*, vol. 19, no. 3, pp. 2269–2272, 2009.
50. J. Bang, S. Kim, J. Y. Jang, Y. J. Hwang, M. Cho, J. Kim, J. T. Lee, M. C. Ahn, S. Lee, and S. Hahn, "Field measurement and analysis of a 3 t 66 mm no-insulation hts nmr magnet with screening current and manufacturing uncertainty considered," *IEEE Transactions on Applied Superconductivity*, vol. 29, no. 5, 2019, Art. no. 4601305.
51. L. Bortot, M. Mentink, C. Petrone, J. Van Nugteren, G. Kirby, M. Pentella, A. Verweij, and S. Schops, "Numerical analysis of the screening current-induced magnetic field in the hts insert dipole magnet feather-m2. 1-2," *Superconductor Science and Technology*, vol. 33, no. 12, 2020, Art. no. 125008.
52. E. Berrospe-Juarez, F. Trillaud, V. M. Zermeño, F. Grilli, H. W. Weijers, and M. D. Bird, "Screening currents and hysteresis losses in the rebco insert of the 32 t all-superconducting magnet using ta homogenous model," *IEEE Transactions on Applied Superconductivity*, vol. 30, no. 4, 2020, Art. no. 4600705.
53. Y. Yan, Y. Li, and T. Qu, "Screening current induced magnetic field and stress in ultra-high-field magnets using rebco coated conductors," *Superconductor Science and Technology*, vol. 35, no. 1, 2021, Art. no. 014003.
54. Y. Sogabe, K. Wakabayashi, and N. Amemiya, "Magnetic field drifts of small hts dipole magnet under repeated excitation," *IEEE Transactions on Applied Superconductivity*, vol. 32, no. 4, 2022, Art. no. 4600305.
55. N. Amemiya, Y. Sogabe, M. Sakashita, Y. Iwata, K. Noda, T. Ogitsu, Y. Ishii, and T. Kurusu, "Magnetisation and field quality of a cosine-theta dipole magnet wound with coated conductors for rotating gantry for hadron cancer therapy," *Superconductor Science and Technology*, vol. 29, no. 2, 2015, Art. no. 024006.
56. P. Fazilleau, F. Borgnolutti, G. Dilasser, and M. Durante, "Screening currents within the eucard hts dipole," *IEEE Transactions on Applied Superconductivity*, vol. 28, no. 4, 2018, Art. no. 4604605.
57. J. Park, G. Kim, G. Hahn, D. Kim, T. Ha, J. Bang, J. Kim, S. Shin, and S. Hahn, "Estimation of screening current induced field on short period planar hts undulator," *IEEE Transactions on Applied Superconductivity*, vol. 33, no. 5, 2023, Art. no. 4100505.
58. K. Zhang and M. Calvi, "Review and prospects of world-wide superconducting undulator development for synchrotrons and fels," *Superconductor Science and Technology*, 2022.
59. S. Hahn, D. K. Park, J. Bascunan, and Y. Iwasa, "Hts pancake coils without turn-to-turn insulation," *IEEE transactions on applied superconductivity*, vol. 21, no. 3, pp. 1592–1595, 2010.
60. J. Liu, Q. Wang, L. Qin, B. Zhou, K. Wang, Y. Wang, L. Wang, Z. Zhang, Y. Dai, H. Liu, *et al.*, "World record 32.35 tesla direct-current magnetic field generated with an all-superconducting magnet," *Supercond. Sci. Technol.*, vol. 33, no. 3, p. 03LT01, 2020.
61. S. Yoon, J. Kim, K. Cheon, H. Lee, S. Hahn, and S.-H. Moon, "26 T 35 mm all-$GdBa_2Cu_3O_{7-x}$ multi-width no-insulation superconducting magnet," *Supercond. Sci. Technol.*, vol. 29, no. 4, p. 04LT04, 2016.
62. L. Shao, X. Zhang, Y. Yan, H. Wang, H. Liu, and T. Qu, "Design of a 20 T class REBCO insert in a 15 T low temperature superconducting magnet," *Electronics*, vol. 10, no. 14, 2021, Art. no. 1741.
63. S. Hahn, K. Kim, K. Kim, X. Hu, T. Painter, I. Dixon, S. Kim, K. R. Bhattarai, S. Noguchi, J. Jaroszynski, *et al.*, "45.5-tesla direct-current magnetic field generated with a high-temperature superconducting magnet," *Nature*, pp. 496–499, 2019.

64. T. Lécrevisse, A. Badel, T. Benkel, X. Chaud, P. Fazilleau, and P. Tixador, "Metal-as-insulation variant of no-insulation hts winding technique: pancake tests under high background magnetic field and high current at 4.2 k," *Superconductor Science and Technology*, vol. 31, no. 5, 2018, Art. no. 055008.
65. J. Kim, Y. Kim, S. Yoon, K. Shin, J. Lee, J. S. Jung, J. T. Lee, J.-G. Kim, D. Kim, J. Yoo, *et al.*, "Design, construction, and operation of an 18 t 70 mm no-insulation (re) ba2cu3o7- x magnet for an axion haloscope experiment," *Rev. Sci. Instrum.*, vol. 91, no. 2, 2020, Art. no. 023314.
66. H. Miyazaki, S. Iwai, Y. Otani, M. Takahashi, T. Tosaka, K. Tasaki, S. Nomura, T. Kurusu, H. Ueda, S. Noguchi, *et al.*, "Design of a conduction-cooled 9.4 T REBCO magnet for whole-body MRI systems," *Supercond. Sci. Technol.*, vol. 29, no. 10, p. 104001, 2016.
67. T. Painter, K. Kim, K. Kim, K. Bhattarai, K. Radcliffe, H. Xinbo, S. Bole, B. Jarvis, I. Dixon, S. Hahn, *et al.*, "Design, construction and operation of a 13 T 52 mm no insulation REBCO insert for a 20T all superconducting user magnet," in *25th International Conference on Magenet Technology Or31-03*, 2017.
68. F. Borgnolutti, A. Badel, T. Benkel, X. Chaud, F. Debray, P. Fazilleau, T. Lecrevisse, and P. Tixador, "Design study of a 10-T REBCO insert solenoid," *IEEE. Trans. Appl. Supercond.*, vol. 26, no. 4, p. 4600405, 2016.
69. J. Y. Jang, S. Yoon, S. Hahn, Y. J. Hwang, J. Kim, K. H. Shin, K. Cheon, K. Kim, S. In, Y.-J. Hong, *et al.*, "Design, construction and 13 K conduction-cooled operation of a 3 T 100 mm stainless steel cladding all-REBCO magnet," *Supercond. Sci. Technol.*, vol. 30, no. 10, p. 105012, 2017.
70. P. C. Michael, D. Park, Y. H. Choi, J. Lee, Y. Li, J. Bascuñán, S. Noguchi, S. Hahn, and Y. Iwasa, "Assembly and test of a 3-nested-coil 800-MHz REBCO insert (H800) for the MIT 1.3 GHz LTS/HTS NMR magnet," *IEEE. Trans. Appl. Supercond.*, vol. 29, no. 5, p. 4300706, 2019.
71. P.-H. Chen, C. Gao, N. Alaniva, S. Björgvinsdóttir, I. G. Pagonakis, M. A. Urban, A. Däpp, R. Gunzenhauser, and A. B. Barnes, "Watch-sized 12 tesla all-high-temperature-superconducting magnet," *Journal of Magnetic Resonance*, p. 107588, 2023.
72. K. L. Kim, S. Yoon, K. Cheon, J. Kim, H. Lee, S. Lee, D. L. Kim, and S. Hahn, "400-mhz/60-mm all-rebco nuclear magnetic resonance magnet: Magnet design," *IEEE Transactions on Applied Superconductivity*, vol. 26, no. 4, 2015, Art. no. 4302604.
73. J.-B. Song, S. Hahn, T. Lécrevisse, J. Voccio, J. Bascuñán, and Y. Iwasa, "Over-current quench test and self-protecting behavior of a 7 t/78 mm multi-width no-insulation rebco magnet at 4.2 k," *Superconductor Science and Technology*, vol. 28, no. 11, 2015, Art. no. 114001.
74. J. Choi, S. Lee, C. K. Lee, D.-G. Kim, M. Park, and K. Sim, "Electromagnetic design of consecutive hts quadrupole magnets for the preseparator at risp," *IEEE Transactions on Applied Superconductivity*, vol. 28, no. 3, 2018, Art. no. 4006806.
75. U. Bong, K. Choi, J. Kim, J. Bang, J. T. Lee, C. Im, J. Park, J. Yoon, G. Kim, H. Ku, *et al.*, "Design, construction, and operation of a 2 t 240 mm conduction-cooled defect-irrelevant winding (re) ba2cu3o7- x magnet," *Review of Scientific Instruments*, vol. 93, no. 7, 2022.
76. Y. Suetomi, T. Yoshida, S. Takahashi, T. Takao, G. Nishijima, H. Kitaguchi, Y. Miyoshi, M. Hamada, K. Saito, R. Piao, *et al.*, "Quench and self-protecting behaviour of an intra-layer no-insulation (lni) rebco coil at 31.4 t," *Superconductor Science and Technology*, vol. 34, no. 6, 2021, Art. no. 064003.
77. D. Park, W. Lee, J. Bascuñán, H. M. Kim, and Y. Iwasa, "A cryogen-free 25-t rebco magnet with the extreme-no-insulation winding technique," *IEEE Transactions on Applied Superconductivity*, vol. 32, no. 6, 2022, Art. no. 4602305.
78. M. Durante, F. Borgnolutti, D. Bouziat, P. Fazilleau, J.-M. Gheller, F. Molinié, and P. De Antoni, "Realization and first test results of the eucard 5.4-t rebco dipole magnet," *IEEE Transactions on Applied Superconductivity*, vol. 28, no. 3, 2018, Art. no. 4203805.
79. J. Bang, J. T. Lee, K. Choi, S. H. Park, U. Bong, S. An, C. Im, J. Park, J. Kim, K. Bhattarai, *et al.*, "A numerical method to calculate spatial harmonic coefficients of magnetic fields generated by screening currents in an hts magnet," *IEEE Transactions on Applied Superconductivity*, vol. 30, no. 4, 2020, Art. no. 4901405.

80. A. Pippard, "The coherence concept in superconductivity," *Physica*, vol. 19, no. 1-12, pp. 765–774, 1953.
81. A. R. Innes and E. Rhoderick, *Introduction to superconductivity*. Pergamon Press, 1980.
82. P. J. Ray, *Structural Investigation of La (2-x) Sr (x) CuO (4+ y): Following a Staging as a Function of Temperature*. Niels Bohr Institute, Copenhagen University, 2015.
83. "Sunam's 2g hts wire architecture," http://www.i-sunam.com/wp/sunam1-2/, 2022, [Online; accessed 13-oct-2023].
84. A. Sundaram, Y. Zhang, A. Knoll, D. Abraimov, P. Brownsey, M. Kasahara, G. Carota, R. Nakasaki, J. Cameron, G. Schwab, *et al.*, "2G HTS wires made on 30 μm thick Hastelloy substrate," *Superconductor Science and Technology*, vol. 29, no. 10, 2016, Art. no. 104007.
85. J. Kim, S. Yoon, K. Cheon, K. H. Shin, S. Hahn, D. L. Kim, S. Lee, H. Lee, and S.-H. Moon, "Effect of resistive metal cladding of HTS tape on the characteristic of no-insulation coil," *IEEE Transactions on Applied Superconductivity*, vol. 26, no. 4, 2016.
86. S. Hahn, Y. Kim, D. Keun Park, K. Kim, J. P. Voccio, J. Bascuñán, and Y. Iwasa, "No-insulation multi-width winding technique for high temperature superconducting magnet," *Applied physics letters*, vol. 103, no. 17, 2013.
87. S. Hahn, J. Song, Y. Kim, T. Lécrevisse, Y. Chu, J. Voccio, J. Bascuñán, and Y. Iwasa, "Construction and test of 7-t/68-mm cold-bore multiwidth no-insulation gdbco magnet," *IEEE Transactions on Applied Superconductivity*, vol. 25, no. 3, 2014, Art. no. 4600405.
88. J. Schwinger, "The quantum correction in the radiation by energetic accelerated electrons," *Proceedings of the National Academy of Sciences*, vol. 40, no. 2, pp. 132–136, 1954.
89. S. Wimbush and N. Strickland, "Critical current characterisation of sunam san04200 2g hts superconducting wire," https://doi.org/10.6084/m9.figshare.5182354.v1, [Online; accessed 13-oct-2023].
90. U. Bong, S. An, J. Park, and S. Hahn, "Numerical studies on mechanical behavior of dry-wound hts racetrack coil," *IEEE Transactions on Applied Superconductivity*, vol. 30, no. 4, 2020, Art. no. 4901805.
91. J. Xia, H. Bai, H. Yong, H. W. Weijers, T. A. Painter, and M. D. Bird, "Stress and strain analysis of a rebco high field coil based on the distribution of shielding current," *Superconductor Science and Technology*, vol. 32, no. 9, 2019, Art. no. 095005.
92. D. Kolb-Bond, M. Bird, I. Dixon, T. Painter, J. Lu, K. Kim, K. Kim, R. Walsh, and F. Grilli, "Screening current rotation effects: Scif and strain in rebco magnets," *Superconductor Science and Technology*, vol. 34, no. 9, 2021, Art. no. 095004.
93. J. Park, J. Bang, U. Bong, J. Kim, D. Abraimov, and S. Hahn, "Parametric study on effect of friction and overbanding in screening current stress of lbc magnet," *IEEE Transactions on Applied Superconductivity*, vol. 31, no. 5, 2021, Art. no. 4603205.
94. O. A. Bauchau and J. I. Craig, *Euler-Bernoulli beam theory*. Springer, 2009.
95. S. cryogenics group, "Srdk-408d2 cold head capacity map," https://www.shicryogenics.com/product/rdk-408d2-4k-cryocooler-series/, [Online; accessed 27-oct-2023].
96. X. Wang, S. Hahn, Y. Kim, J. Bascuñán, J. Voccio, H. Lee, and Y. Iwasa, "Turn-to-turn contact characteristics for an equivalent circuit model of no-insulation rebco pancake coil," *Superconductor Science and Technology*, vol. 26, no. 3, 2013, Art. no. 035012.
97. Y. Wang, W. K. Chan, and J. Schwartz, "Self-protection mechanisms in no-insulation (re) ba2cu3ox high temperature superconductor pancake coils," *Superconductor Science and Technology*, vol. 29, no. 4, p. 045007, 2016.
98. P. Bruzzone, "The index n of the voltage–current curve, in the characterization and specification of technical superconductors," *Physica C: Superconductivity*, vol. 401, no. 1-4, pp. 7–14, 2004.
99. M. Gyimesi and D. Ostergaard, "Inductance computation by incremental finite element analysis," *IEEE transactions on magnetics*, vol. 35, no. 3, pp. 1119–1122, 1999.
100. H. L. Quach, J. H. Kim, Y. S. Chae, J. H. Moon, J. H. Ko, S. H. Chae, J. H. Ahn, E.-H. Kim, S. H. Kim, Y. G. Jin, *et al.*, "Analysis on electrical and thermal characteristics of a no–insulation hts coil considering heat generation in steady and transient states," *IEEE Transactions on Applied Superconductivity*, vol. 29, no. 5, 2019, Art. no. 4701506.

101. D. G. Yang, J.-B. Song, Y. H. Choi, S.-G. Kim, Y. S. Choi, and H. Lee, "A study on electrical characteristics of multilayered metallic-insulation coils," *IEEE Transactions on Applied Superconductivity*, vol. 27, no. 4, 2017, Art. no. 7700206.
102. K. L. Kim, S. Hahn, Y. Kim, D. G. Yang, J.-B. Song, J. Bascuñán, H. Lee, and Y. Iwasa, "Effect of winding tension on electrical behaviors of a no-insulation rebco pancake coil," *IEEE transactions on applied superconductivity*, vol. 24, no. 3, 2013, Art. no. 4600605.
103. T. S. Lee, Y. J. Hwang, J. Lee, W. S. Lee, J. Kim, S. H. Song, M. C. Ahn, and T. K. Ko, "The effects of co-wound kapton, stainless steel and copper, in comparison with no insulation, on the time constant and stability of gdbco pancake coils," *Superconductor Science and Technology*, vol. 27, no. 6, 2014, Art. no. 065018.
104. M. Bonura, C. Barth, A. Joudrier, J. F. Troitino, A. Fete, and C. Senatore, "Systematic study of the contact resistance between rebco tapes: pressure dependence in the case of no-insulation, metal co-winding and metal-insulation," *IEEE Transactions on Applied Superconductivity*, vol. 29, no. 5, 2019, Art. no. 6600305.
105. M.-H. Sohn, K. Sim, B. Eom, H.-S. Ha, H.-Y. Kim, and K. Seong, "Controllability of the contact resistance of 2g hts coil with metal insulation," *IEEE Transactions on Applied Superconductivity*, vol. 28, no. 3, 2018, Art. no. 4602705.
106. R. Wang, G. Ma, P. Zhou, T. Gong, and S. Li, "Thermo-electromagnetic modeling of coated superconductor coils with metal insulation," *Superconductor Science and Technology*, vol. 34, no. 11, 2021, Art. no. 115017.
107. S. Xue, J. Kwon, Y. Guo, T. Garg, M. Sumption, and E. Collings, "Compressive stress-strain behavior of rebco coated conductors and cables," *IEEE Trans. Appl. Supercond.*, vol. 33, no. 5, 2023, Art. no. 4800706.
108. Y. Yan, J. Kim, U. Bong, W. Jang, G. Kim, J. Park, and S. Hahn, "Measurement and analysis of winding stresses in dry-wound pancake coils considering nonlinear compressive behaviors," *Superconductor Science and Technology*, vol. 36, no. 11, 2023, Art. no. 115019.
109. W. Edwards and G. Boulton, "The mystery of coil winding," in *2001 Iron and Steel Exposition and AISE Annual Convention*, 2001, p. 2001.
110. J. A. Greenwood and J. Tripp, "The contact of two nominally flat rough surfaces," *Proceedings of the institution of mechanical engineers*, vol. 185, no. 1, pp. 625–633, 1970.
111. J. A. Greenwood and J. P. Williamson, "Contact of nominally flat surfaces," *Proceedings of the royal society of London. Series A. Mathematical and physical sciences*, vol. 295, no. 1442, pp. 300–319, 1966.
112. J. Good and Y. Xu, "Computing wound roll stresses based upon web surface characteristics," 1993.
113. J. Park, J. Kim, Y. Yan, and S. Hahn, "Diameter change of dry wound rebco solenoid pancake coil by winding tension," *IEEE Transactions on Applied Superconductivity*, vol. Submitted, 2023.
114. V. Arp, "Stresses in superconducting solenoids," *J. Appl. Phys*, vol. 48, no. 5, pp. 2026–2036, 1977.
115. W. Markiewicz and C. Swenson, "Winding strain analysis for ybco coated conductors," *Superconductor Science and Technology*, vol. 23, no. 4, 2010, Art. no. 045017.
116. S. Ito, H. Hashizume, N. Yanagi, and H. Tamura, "Bridge-type mechanical lap joint of hts stars conductors using an integrated joint piece," *Fusion Eng. Des.*, vol. 146, pp. 590–593, 2019.
117. X. Hu, M. Small, K. Kim, K. Kim, K. Bhattarai, A. Polyanskii, K. Radcliff, J. Jaroszynski, U. Bong, J. H. Park, *et al.*, "Analyses of the plastic deformation of coated conductors deconstructed from ultra-high field test coils," *Supercond. Sci. Technol.*, vol. 33, no. 9, 2020, Art. no. 095012.
118. B. Shen, F. Grilli, and T. Coombs, "Review of the AC loss computation for HTS using H formulation," *Supercond. Sci. Technol.*, vol. 33, no. 3, 2020, Art. no. 033002.
119. D. N. Nguyen, S. P. Ashworth, J. O. Willis, F. Sirois, and F. Grilli, "A new finite-element method simulation model for computing AC loss in roll assisted biaxially textured substrate YBCO tapes," *Supercond. Sci. Technol.*, vol. 23, no. 2, 2009, Art. no. 025001.

120. M. Niu, H. Yong, J. Xia, and Y. Zhou, "The effects of ferromagnetic disks on AC losses in HTS pancake coils with nonmagnetic and magnetic substrates," *J. Supercond. Nov. Magn.*, vol. 32, no. 3, pp. 499–510, 2019.
121. V. M. Zermeno, F. Grilli, and F. Sirois, "A full 3d time-dependent electromagnetic model for roebel cables," *Superconductor Science and Technology*, vol. 26, no. 5, 2013, Art. no. 052001.
122. B. Shen, X. Chen, L. Fu, S. Jiang, H. Gou, J. Sheng, Z. Huang, W. Wang, Y. Zhai, Y. Yuan, *et al.*, "Superconducting conductor on round core (corc) cables: 2d or 3d modeling?" *IEEE Transactions on Applied Superconductivity*, vol. 31, no. 8, 2021, Art. no. 4803505.
123. M. Zhang and T. Coombs, "3d modeling of high-tc superconductors by finite element software," *Superconductor science and technology*, vol. 25, no. 1, 2011, Art. no. 015009.
124. L. Wang and Y. Chen, "Hysteresis loss calculations of 2d and 3d large-scale hts coil models: from multi-scale method to multi-dimension method," *Superconductor Science and Technology*, vol. 34, no. 5, p. 055004, 2021.
125. J. Babiskin, "Oscillatory galvanomagnetic properties of bismuth single crystals in longitudinal magnetic fields," *Physical Review*, vol. 107, no. 4, p. 981, 1957.
126. S. Kawabata, "Magnetic permeability of the iron yoke in high field superconducting magnets," *Nuclear Instruments and Methods in Physics Research Section A: Accelerators, Spectrometers, Detectors and Associated Equipment*, vol. 329, no. 1-2, pp. 1–8, 1993.
127. F. Bertinelli, S. Comel, P. Harlet, G. Peiro, A. Russo, and A. Taquet, "Production of low-carbon magnetic steel for the lhc superconducting dipole and quadrupole magnets," *IEEE transactions on applied superconductivity*, vol. 16, no. 2, pp. 1777–1781, 2006.
128. X. Lv, L. Sun, and L. Sun, "Study on magnetic properties of ferromagnetic materials in cryogenic conditions," *Cryogenics*, vol. 130, p. 103643, 2023.
129. A. G. Pronto, M. V. Neves, and A. L. Rodrigues, "Measurement and separation of magnetic losses at room and cryogenic temperature for three types of steels used in hts transformers," *Journal of superconductivity and novel magnetism*, vol. 24, pp. 981–985, 2011.

MIX
Papier aus verantwortungsvollen Quellen
Paper from responsible sources
FSC® C105338

If you have any concerns about our products,
you can contact us on
ProductSafety@springernature.com

In case Publisher is established outside the EU,
the EU authorized representative is:
Springer Nature Customer Service Center GmbH
Europaplatz 3, 69115 Heidelberg, Germany

Printed by Libri Plureos GmbH
in Hamburg, Germany